The
Einstein
Connection

Other books by D. W. Kreger,
published by Windham Everitt Publishing:

The Secret Tao: *Uncovering the Hidden History
and Meaning of Lao Tzu.*

2012 & The Mayan Prophecy of Doom:
*The Definitive Guide to the Mythology and
Science behind the 2012 Prophecies*

The Tao of Yoda: *Based upon the Tao Te Ching
by Lao Tzu*

The
Einstein
Connection

Ancient Myths & Scientific Theories
of a Recurring Global Cataclysm

AND THEIR MYSTERIOUS COUNNECTION
TO ALBERT EINSTEIN

D. W. KREGER

Windham Everitt

Cataloging-in-Publication Data on file

IBSN: 978-0-9833099-3-2

Windham Everitt Publishing
P. O. Box 900922
Palmdale, CA 93590

www.windhameveritt.com

10 9 8 7 6 5 4 3 2 1

Printed in the United States of America

For my wife Jamielly, with all my love.

CONTENTS

PREFACE

If you've read my book, "2012 & The Mayan Prophecy of Doom", you may notice that much of it is similar if not identical to this one. "The Einstein Connection" is really, to a great extent, a second edition of my book on the 2012 prophecy. Why publish a second edition? After all, 2012 has come and gone without incident, as I predicted. So you might think that would be the end of it. The problem is that the scientific evidence, while not supporting the idea of a major cataclysm in 2012, did predict a global cataclysm in the future, possibly in our lifetime. Also, much of the previous book had little to do with the Mayans or their prophecies, and more to do with the scientific research on this topic.

In my previous book, I started with the Mayan prophecy but I quickly looked at other myths of apocalyptic doom from other ancient cultures. I found that many other ancient myths from around the world recorded a pattern of similar cataclysmic events in our past and predicted that it will happen again in the future. I then searched for some scientific theory that might explain how such a recurring natural cataclysm might actually happen. Finally, I researched any evidence that might confirm or disprove both the mythology and the scientific theories I had found.

What I discovered is that our planet is indeed plagued by a recurrent global cataclysm of biblical proportions. I found that the Earth has been nearly destroyed and humanity nearly wiped out numerous times in the near prehistory. And, according to our best scientific research, it will happen again, possibly very soon. In fact the only thing that the scientific evidence did not support was the predicted date of an apocalypse in 2012.

It is important to present my research quite independent of all the hoopla surrounding the Mayan 2012 prophecy. I have revised my previous book, taken out some detailed analysis of Mayan calendars and such, and added details that are more relevant to the topic. Primarily, I added details concerning an unlikely doomsday theorist named Albert Einstein.

One interesting and important detail that I think got lost in my last book was the contributions of Albert Einstein to this line of research. Many people do not realize that Einstein was instrumental in collaborating with scientists who were working on theories to explain the end of the last ice age, and the extinction level event at the end of the Pleistocene. Einstein also collaborated with other scientists on magnetic pole shifts and their connection to leaps in human evolution. All of these are aspects of modern doomsday scenarios, but to Einstein they were just good science. This is the "Einstein connection" that this edition is named for. As far as we know

Einstein had no knowledge of or was unconcerned with any ancient myths predicting "the end of the world." So, in our rush to discard anything related to the 2012 prophecies, Einstein's research on earth science and evolution regarding a potential looming global disaster, has also been discarded.

This book is an effort to keep the research of Albert Einstein, Charles Hapgood, Oliver Reiser, and many other distinguished scientists, from being lumped together with 2012, Y2K, or any other failed predictions of the past.

These are serious scientists concerned about a very real and reoccurring pattern of global cataclysms that definitely have plagued the earth for millions of years, and most assuredly will strike again.

Finally I want to thank all the thousands of people who have supported my work, and given me positive feedback on my research. It is always gratifying to know that one's work is respected and appreciated.

D. W. Kreger
September 20, 2013

Part I

MYTHS & PROPHECIES
OF DOOM

i

THE EINSTEIN CONNECTION

"A person who never made a mistake never tried anything new."

-Albert Einstein.

The study of doomsday prophecies and global cataclysmic scenarios may sound crazy to some people. It sounded pretty crazy to me a few years ago, but that was before I began seriously researching this topic. What I found both interested and alarmed me.

I discovered that extinction level events, or at least extreme global events, have occurred repeatedly and regularly for millions of years. The Earth has been very nearly destroyed a number of times. Ice Ages, asteroid and comet impacts, and some events we don't fully understand yet, have nearly wiped out humanity a number of times. And, it is inevitable it will happen again. The only question is when. And, if we can predict what will happen and when, we have a better chance of minimizing the tragedy and loss of life. That's what makes this research so interesting and important.

How did a respectable researcher in the fields of psychology and archaeology get involved in studying doomsday scenarios? One day in 2009, a troubled student came to me complaining of panic attacks because of his fear of "the end of the world" after he'd seen a documentary on cable about the Mayan doomsday prophecy. The prophecy, if you don't already know, predicted a major cataclysmic event, some say the end of the world, on December 21, 2012. He asked me about the prophecy, and whether there was any truth to it. I said that I doubted it. But, in truth I knew very little about it. So, I decided I should do a little research.

That research lead to over 2 years of serious investigation, resulting in the book that you are now reading. Of course there was no global catastrophe in 2012, besides the usual collection of regional natural disasters. What I discovered, however, did not comfort me, even after the year 2012 passed without incident.

I began my quest by researching the Mayan Prophecies. That led to further research on Mayan culture, history, beliefs, astronomy and so on. I also read other writers on this topic from John Major Jenkins[1] to Jose Argüelés[2], as well as Mayan scholars skeptical of the prophecies such as David Stuart[3] at the University of Texas, and Anthony Aveni[4] at Colgate University. The more I researched the topic, however, the more interested I became. I eventually went on to research scientific theories about global cataclysms and how they might occur.

One thing in particular caught my attention. More

than once, while reading about some obscure theorist's ideas about a periodically recurring global cataclysm, I found that the theorist had been corresponding with Albert Einstein, and that Einstein had contributed to their ideas and research. This was not an isolated event. Einstein was seriously interested in earth science and human evolution, and helped to shape some of the theories on this topic. Over time, I came to refer to these theories and this topic of study in general as "The Einstein Connection", hence the name of the book.

Using the Scientific Method

To see if there was any truth to these Mayan prophecies, I decided to employ the scientific method. That is, I proposed something of a hypothesis testing approach. This is how I was trained as a behavioral scientist, and it is how most serious scientific investigations are conducted. This meant expanding my research to include not just the myths, but whatever scientific theories and evidence might exist to support them.

Step 1: Identify all the salient features of the ancient mythology and prophecies in the surviving ancient Mayan texts regarding global cataclysms.

Step 2: Look for corroborating mythology and prophecies from other ancient cultures around the world. After all, were the Mayans so much more advanced than the ancient Greeks, Egyptians, or Sumerians?

Step 3: Edit our list of myths and prophecies regarding global cataclysms to only those features found in both Mayan and non-Mayan cultures. If there was a myth of a disaster recorded by one culture but not another, maybe it wasn't such a global disaster after all.

Step 4: Look for any scientific theory that might possibly explain the specific details recorded in the myths and prophecies on our list.

Step 5: Finally, in our hypothesis-testing phase, we look at what if any evidence exists to support the myths of ancient cataclysms, or to support scientific theories about how such cataclysms could occur. (To be precise we'd want to prove the null hypothesis, if we're going to strictly follow the scientific method.)

I did all that, I completed all of the above steps. And, guess what? I was, indeed, able to identify a list of corroborated myths and prophecies from ancient cultures around the world. I also found a few, not very well known theories which, taken together, could explain exactly the same type of events listed in the myths and prophecies. I even found evidence to support these myths, prophecies, and scientific theories, except for one little detail. The timing was off.

There definitely is evidence of periodic cataclysmic events nearly killing off Humanity in the not-too distant prehistoric past.[5] And there are scientific theories, though controversial, which explain how these events could happen.[6] But even though I could find evidence to support the myths and theories, none of my research supported the date of 2012. I was forced to admit that, while a global cataclysm on Earth may occur in our lifetime, there was no evidence that it would happen in 2012. We merely had the Mayan Priest's prophecies and even that was iffy. Many Mayan scholars suggested that the prophecies were not being interpreted correctly.

So, I had predicted that most likely nothing would happen in 2012, and I was right; it didn't. It's not so much that I knew nothing would happen, but I just didn't find any evidence to support the idea that it definitely would happen in the year 2012, much less an exact day in 2012. I didn't get much fanfare for the accuracy of my prediction. Skeptics of all sorts predicted the same outcome. But, of all the skeptics, I'm the only one who was open minded enough to give the prophecies the benefit of the doubt. And I don't know of anyone else who actually used the scientific method to research the probability of a cataclysm in 2012.

I'll admit, I was a little disappointed. Not that I would like to see a global cataclysm, but I thought that something auspicious might occur in December of 2012. Maybe we might find a new star, or planet, or we might have a near miss with a large asteroid. Actually such discoveries were made in 2012, but not necessarily on December 21, 2012, and none were very meaningful or alarming finds. As of today, the lost planet of Nibiru still has not been spotted.[7]

So much for the importance of the year 2012. But, that doesn't mean we're completely out of danger.

As I mentioned earlier, the fact that nothing happened did not make me relieved in the slightest. First, I predicted that nothing would happen in 2012, so that was not a surprise. Second, we are still left with the fact that the world has endured periodic global cataclysms many times before and there is no doubt that it will happen again, as I'll show later. And, there are serious scientific theories to explain what might happen and how, and some evidence to back up those theories. The only thing we don't have is what the Mayan prophecy offered, the precise date of our doomsday prediction. But, then again, the scientific research isn't complete yet. In fact, it's barely begun. And, there are reasons to be encouraged.

The Einstein Connection

That one little coincidence, the one I call "the Einstein connection," proved to be even more coincidental than I had earlier thought. At a certain point it was sort of bizarre. As I said, the connection with Einstein came up several times. The funny thing was that, in the end, when we eliminated any scientific theory that did not specifically explain the events described in the ancient mythology, only three theories were left, and two out of three had direct connections with Einstein.

Einstein actually consulted or collaborated on two of these theories. And, the third theory concerned binary star systems and the astrophysics of our solar system. Ironically I found no direct connection between Einstein and that theory, but as he was the most eminent astrophysicist of his time, he almost certainly was aware of it, and probably had some opinion on it. And, I have found no evidence that he disagreed with the theory. So, of the three theories, which could explain the events described in ancient mythology, all three were directly or tangentially connected to Albert Einstein.

Maybe this means nothing. It could easily just be a coincidence. But, thinking that Albert Einstein vetted these scientific theories somehow adds credence to them. Perhaps it might prompt the scientific community take a second look at them, and to redouble our efforts to understand how these cataclysms occurred and when to expect the next one.

As I said, all these theories are very controversial, however, Einstein himself was controversial.[8] His quotes tell us how he approached science. He said "The only thing that interferes with my learning is my education." Even more important, Einstein said that "Imagination is more important than knowledge. For knowledge is limited to all we now know and understand, while imagination embraces the entire world, and all there will ever be to know and understand."[9]

More than anything else Einstein was curious. He said that "I have no special talent. I am only passionately curious."

From the above quotes it's clear that Einstein was more concerned with the possibilities of what might be, than the certainty of what we know. Of course, to a curious mind what we already know is boring. The really interesting thing is what we don't know yet. I think this attitude stemmed from his creativity and his playful spirit. It was this playful spirit that that separated him from other scientists of his time. He was once quoted as saying "Creativity is intelligence having fun."

The scientific theories he worked on regarding global cataclysms were so monumental and important that anyone in their right mind would be curious if not downright concerned. Yet, many scientists have discarded these theories as flawed. If we know that violent, global cataclysms regularly visit the Earth, and there are theories that might explain these events, then every scientist on Earth should be researching and trying to refine these theories, not discard them because they're flawed. One can't help but wonder why more scientists today aren't more curious. And what could be more important than quite literally saving the world? You'd think that would be an engrossing problem to work on. Yet, the scientific theories presented in this book, while receiving some praise and attention at first, have been almost forgotten. Perhaps it is as Einstein once said "Intellectuals solve problems, geniuses prevent them."

Einstein was open-minded when it came to science but skeptical when it came to people. He once said: "If my theory of relativity is proven successful, Germany will claim

me as a German, and France will declare that I am a citizen
of the world. Should my theory prove untrue, France will say
that I am German, and Germany will declare that I am a Jew."
He probably would have understood and expected opposition
from mainstream science. He is quoted as saying that "Great
spirits have always encountered violent opposition from
mediocre minds." So, he probably would not be surprised by
those critical of this research. He famously said "Two things
are infinite: the Universe, and human stupidity, and I'm not
sure about the Universe."

The Jigsaw Puzzle of Theories

One thing I uncovered in my research on those three
cataclysmic theories is that the gaps in one theory often are
filled in by one of the other theories. It's as if several different
theorists were each working on one part of the problem, but
no one had put all the pieces together. When you look at all
three of these theories they fit together like a jigsaw puzzle.

This brings up an interesting question. The only person to
know of and/or work on all three of these theories was Albert
Einstein. There is no evidence that any of these theorists
communicated or collaborated with each other, yet each was
in communication and consulting with Einstein. But, there is
no evidence that Einstein saw these theories fitting together,
completing gaps in each other, and becoming one all-inclusive
theoretical model. Or, did he?

Perhaps Einstein did see but did not comment on the complementary nature of these different theories. Who knows what might have happened if he had lived another 15 or 20 years? He might well have made those connections, and integrated these different theories into a larger cataclysmic scenario. Einstein often joked that he wasn't smarter, just more persistent. He said "I think and think for months and years, ninety-nine times the conclusion is false. The hundredth time I am right."

What is important about the "Einstein Connection" is not really Einstein himself or the focus of his research, which was obviously concentrated on astrophysics and relativity, but rather this interesting line of research that he happened to have had a minor role in. I call this line of research *The Einstein Connection*, for lack of a better term, because he is the only common denominator among this otherwise unusual group of seemingly unrelated scientific theories.

The important features of this line of research is a) there are a handful of credible scientific theories that point to the existence of a major, recurring, global cataclysm, b) this recurring event might well be a factor in propelling human evolution, c) if true, it would make us also re-think the astrophysics of our solar system forever, d) the mainstream scientific community has largely ignored these theories, and e) this could have dire consequences for the future of Earth and humanity.

What we need now is another Einstein, someone capable of thinking outside the box, and not getting so mired in data and procedures that they lose sight of the big picture and the looming threat that clearly exists. As Einstein said "We can't solve problems by using the same kind of thinking we used when we created them." So many scientists today are not interested in that which is mysterious, only that which is explainable.[10] But as Einstein said: "The most beautiful thing we can experience is the mysterious. It is the source of all true art and science."

What follows is the distillation of my research on ancient mythology and prophecies, from around the world. And, in the next sections, I'll present scientific theories to account for this phenomena, and the evidence I uncovered regarding each. Hopefully this book will inspire a new generation of curious scientists because, if so, one of them just might end up saving the world.

"Remember today, for it is the beginning of always.
Today marks the start of a brave new future
filled with all your dreams can hold. Think truly to
the future and make those dreams come true"

-Albert Einstein

2

MYTHS & PROPHECIES OF THE ANCIENT MAYA

What exactly are we talking about when we refer to the Mayan Prophecies? There are actually only 5 Mayan books that remain after the Spanish conquest and the subsequent massive book burning by the conquistadors. If the other thousands of Mayan books were anything like the 5 remaining books, this was an amazing and a very advanced civilization. The surviving books only cover a few topics but they are very impressive. They involve a great deal of advanced mathematics and astronomy with precise predictions of solar and lunar eclipses, the phases of Venus, and a very advanced calendar system that accurately measured the year within a few thousandths of a day. Their scientific knowledge on these topics was much more advanced than that of the Europeans at that time.

In addition to their impressive scientific writings, the books also contained the history and mythology of the Mayan people. One book, the Popol Vuh (The Community Book), contained an elaborate creation myth.[1] Though, in principle, it is not unlike the book of Genesis in the Bible, this creation story is very involved and the narrative quite different in many respects. At least a couple of the books, written by the Chilam Balam (Jaguar Priests or "great priests"), seem to include prophecies, as well as history and mythology.[2] The Chilam Balam books talk of the conquest by the Spaniards, including the prophecies related to them, the calendar omens –not unlike our own astrology– and some details about the time of the conquest. The books also contain prophecies for the future of the Mayan People.

For our purposes, it is most important to highlight 3 aspects of these books. First, the Mayan Calendar, from this we get the Long Count, a period of over 5000 years, the end of which they believed would be the end of this epoch of humanity. Secondly, we need to study their creation mythology. They believed that there have been several different creation events. Each time, the gods were unhappy and eventually destroyed what they had made. Then, they would re-create humanity once again, each time a little better than before. We are in the 4th or 5th such epoch now. Lastly, we need to study the prophecies for the future. They most definitely do prophesy great destruction and suffering in the future. But, they also predict some wonderful things in store for us.

Popol Vuh

Of the five books that we have left, from the vast and ancient Mayan libraries, one stands out as the Bible of the Mayans. It is the Popol Vuh.[3] Loosely translated as the "Community Book", it means that it is "The Book of the People" and it truly is the book of the Maya, their history, mythology, and their story as a people. It tells their history from the beginning of Creation. It tells of the different tribes of the Maya and how each tribe came about. It tells of their heroic feats in the past, and their struggles with gods, and with the underworld. If you want to understand the philosophical and mythological foundation of the Maya people, this is the book to read.

The Mayan story of Creation is very similar to other Mesoamerican stories, such as the Aztecs, in that it involves a number of different attempts by the gods to create people. Each time they were disappointed by their creation, and they would destroy the population and start again.

There are several principle gods involved. There is Tepeu and Gucumatz, also translated as the "Sovereign" and the "Quetzal Serpent," which means feathered serpent. They were considered great sages and thinkers, and they seem to be observing, consulting, and presiding over the Creation. Then, there is something they call the Framer and the Shaper, also translated as The Creator and The Maker, and "She Who Has Borne Children" and "He Who Has Begotten Sons", also known as the "Forefathers," and they really seem to be co-creators of the world. Finally there is the "Heart of Sky" (or the Heart of Heaven), and he is a three part god, Thunderbolt Huracan, Youngest Thunderbolt, and Sudden Thunderbolt. Heart of Sky seems to be a very powerful one, among the gods. He is joined by Heart of Earth.

The First Age

"This is the account of how all was in suspense, all calm, in silence; motionless, still, and the expanse of the sky was empty." So begins the account of Creation in the Popol Vuh.

All the gods convened. "Then came the word," according to the text. Tepeu and Gucumatz talked together and they reached a decision. They arranged for the germination of Creation. Heart of Sky and Heart of Earth joined them and they created the sky and the earth, as separate out of the vast, still waters and emptiness.

And She Who Has Borne Children and He Who Has Begotten Sons said: "Shall it be merely solitary, merely silent beneath the trees and the brushes?" And so they created life. "Then they made the small wild animals, the guardians of the woods, the spirits of the mountains, the deer, the birds, pumas, jaguars, serpents, snakes, vipers, guardians of the thickets."

Once the creation of all of the four-footed animals and birds were finished the Framer and the Shaper said to the animals, "Speak, then, our names, praise us, your mother, your father. Invoke then Thunderbolt Huracan, Youngest Thunderbolt, and Sudden Thunderbolt, the Heart of Sky, the Heart of Earth, the Framer, the Shaper, She Who Has Borne Children, and He Who Has Begotten Sons; speak, invoke us, adore us." But the animals could not. They could only hiss and scream and cackle. They were unable to make words like humans.

When the Framer and the Shaper saw that it was impossible for them to talk, they said, "This is not well." They then condemned them to their current station in life. Their homes were taken away and they were made to lie down in the woods and ravines, and they were condemned to have their flesh torn to pieces, to be killed and eaten by each other. "Accept your destiny. …So shall it be. This shall be your lot."

The Second Age

This time, the Framer and the Shaper had a better idea of how to create people. "Let us try to make obedient, respectful beings who will nourish and sustain us" with their words. This time they formed mud from the earth to create human looking people. But immediately they could see this did not turn out well. The flesh was made of mud. "It melted away, it was soft, did not move, had no strength, it fell down, it was limp, it could not move its head, its face fell to one side, its sight was blurred." The Framer and the Shaper said, let us try again, and they broke up and destroyed their creation.

The Third Age

The Framer and Shaper said: "What shall we do to perfect it, in order that our worshipers, our invoker, will be successful?" They turned to divination. They spoke to Huracan, Tepeu and Gucumatz, they spoke to the soothsayer, to the Grandmother of the Day and the Grandmother of the Dawn. They asked if carving humans out of wood could turn out well. After performing a ceremony, in which they cast grains of maize and tz'ite, the diviners spoke: "May these effigies of wood come out well. May they speak. May they communicate there upon the face of the earth. May it be so." Straightaway effigies of humans were made of wood. The wooden people soon "began to multiply, bearing daughters and sons. Nevertheless, they still did not possess their hearts nor their minds."

They did not remember their Framer or their Shaper. They started to talk, but soon their "faces were all dried up. Their legs and arms were not filled out. They had no blood or blood flow." Worst of all, they were not capable of understanding or appreciating their creators, let alone worship and sustain them with their words of praise. These were the first numerous people who populated the earth, but they too had to be destroyed.

"Then came the end of the effigies carved of wood, for they were ruined, crushed, and killed. A flood was planned by Heart of Sky that came down upon the head of the effigies carved of wood." The Popol Vuh gives more description of this destruction than of any other. It began with a flood but ended with carnage. And, the details were quite gruesome. "There came the ones called Chiselers of Faces who gouged out their eyes. There came the Death Knives, which cut off their heads. There came Crouching Jaguar, who ate their flesh. There came Striking Jaguar who struck them. They smashed their bones and their tendons."

Then an interesting reversal of fortune occurred where the people who kept dogs were mistreated by the dogs, and the animals that they ate, ate them. The animals talked to the people saying "Pain you have caused us. You ate us. Therefore it will be you that we will eat now." Then the dogs said to them: "Why didn't you give us food? All we did was look at you, and you chased us off. You raised sticks against us to beat us while you ate. This day, therefore, you shall try the teeth that are in our mouths. We shall eat you."

The text includes the details of what appears to be a type of natural disaster. "Thus they were killed in the flood. There came a great resin down from the sky. ... Thus they caused the face of the earth to be darkened, and there fell a black rain. The rain fell both day and night."

Finally, it says that "the spider monkeys that are in the forest today are descendents of these people. This was their heritage because their flesh was merely wood when it was created by the Framer and the Shaper. Therefore, the spider monkeys appear like people, descendents of one generation of framed and shaped people. But they were only effigies."

The Fourth Age?

This part of the Popol Vuh gets confusing. It doesn't follow a clear narrative in the Western tradition of story telling. David Stuart seems to suggest that the maize ceremony was one attempt at humanity, like a Third Age, and when that failed, then the Fourth Age of wooden people were created.[4] That would fit neatly with Aztec mythology, which parallels this story pretty closely. But, it seems to me that the maize ceremony was only part of the Third Age, in which the diviners sought to insure the success of the wooden people. At any rate, a fourth segment does exist in the Popol Vuh story, though it doesn't really follow the same structure as the other previous attempts to create humanity.

In this segment of the Popol Vuh, we are told that in the darkness of earth, a demigod "puffed himself up" and he was named Seven Macaw. He set himself up as lord of the flooded people, who were still living in darkness. He pretended to be the light of the sky. And, the people were deceived.

Then the story takes another long and involved twist. It tells the story of two pairs of hero twins. The first pair are One Hunahpu and Seven Hunahpu, and they are the children of the Creators. So, it is not clear if they are gods, or yet another attempt to create human life on earth, as in the previous cycles of creation and destruction. It almost reads like it is an unrelated story that was inserted into the previous narrative, but doesn't really belong here.

The hero twins, One Hunahpu and Seven Hunahpu, are lured to Xibalba, the Mayan underworld, (literally "place of fear") by the gods of the underworld on the pretense of a challenge match

on the ball court. The twins are killed, and their flesh is hung out
to dry. One day, Lady Blood, daughter of the Lord of Xibalba, was
walking past the head of One Hunahpu, which was placed on a
tree. The head spoke to her and said "put out your hand." She did
so, and the head of One Hunahpu spit into her palm, and thus she
was impregnated with twins of her own. This, the second set of
hero twins were Hunahpu and Xbalanque. Literally, the twins were
the offspring of both gods and devils, the offspring equally of both
heaven and hell.

The hero twins faced many more challenges. First they
confront and defeat Seven Macaw, the demigod who pretended to
be the light of the sky. Next, they too were challenged by the lords
of Xibalba, and they were successful. This then, eventually leads to
the next incarnation of humanity.

The Fifth Age

Finally, in the last part of the Popol Vuh, it seems as if we are
rejoining the story from before the introduction of the hero
twins. In this section, She Who Has Borne Children and He Who
Has Begotten Sons, the Framer and the Shaper, and Tepeu and
Gucumatz, decide once again to make another version of humanity.
This time they make people out of corn meal. They grind together
yellow maize and white maize. So, "food entered their flesh, along
with water to give them strength." And this was to be the current
incarnation of humanity.

"Thus their frame and shape were given expression by our
first Mother and our first Father. ...And so there were four who
were made, and mere food was their flesh." It goes on to give their
names. "These are the names of the first people who were framed

and shaped: the first person was Balam Quitze, the second was
Balam Acab, the third was Mahucutah, and the fourth was Iqui
Balam. These then were the names of our first mothers and fathers."

This also appears to be the first dawn since the darkness
that followed the flood, that destroyed the wood people. The people
eagerly awaited the coming of the Sun. "They fixed their eyes firmly
on their dawn, looking there to the east. They watched closely
for the Morning Star (Venus), the Great Star that gives its light at
the birth of the Sun. They looked to the womb of the sky and the
womb of the earth."

At first, it says that the people were perfect, too perfect.
"Perfect was their sight, and perfect was their knowledge of
everything beneath the sky." These people were truly grateful to
their creators for their life, but the gods were still not pleased. The
gods felt somewhat disturbed by these perfect people, since they
were almost gods in their own right. The gods agreed that the
people's vision and knowledge needed to be reduced just a bit, that
they may be more humbled before the Creators. So the people's
eyes were blurred by Heart of Sky. "They were blinded like breath
upon the face of a mirror." And so too was their wisdom lost, there
in their beginning.

Feeling in need of protection, without their perfect vision
or wisdom, they set out for a city across the sea to the East. It
was a fabled and famous city where they hoped they would find
protection and counsel as well as the authority to rule. The city
was called Tulan Zuyva, and it was a great city, on an island, and
surrounded by a channel of water. From the ruler of Tulan, whose
name was Nacxit, they received the instruments of power and the
authority to rule. And, so they returned to their land to rule. They

moved from place to place at first. They had children, and the 4 original people and their children eventually became the tribes of the Mayan empire.

"There had begun the fifth generation of men." They go on to list nine families, "and having ended the dispute over the sisters and the daughters, they carried out the plan of dividing the kingdom into twenty four great houses, as they did." This then is the history of the tribal structure of the Maya people. "The nine lords of Cavec formed the nine families; the lords of Nihaib formed another nine; the lords of Ahau-Quiche formed another four; and the lords of Zaquic formed another two families."

From here they go on to tell the Mayan recorded history, of the settlement of cities such as Cumarcah, and of great lords and kings who lived in those times. This is where the mythology slowly blends into actual history. We don't know how accurate they are, but the end of the Popol Vuh seems to set the stage for the actual historical people that we know as Maya, specifically the Quiche Maya. This is important because, as Stuart points out, this might well be just one regional "Community Book" and there may have been others that differ somewhat. He points out that there are mythological themes found in ancient pottery that appear to have nothing to do with the Popol Vuh.[5] While the Popol Vuh is important because it is the only surviving "Community Book" telling the Mayan mythology, the story may have varied a little from one ethnic group to another. And we don't know how much of the original mythology was either lost or garbled in this version of their creation mythology.

Analysis of the Popol Vuh

First of all, I'd like to point out that I worked from two different translations of the Popol Vuh.[6] In order to put together the most cogent story possible, I drew from the best of both translations, quoting from each of them. One translation used the Mayan names for the gods, such as Tepeu and Gucumatz; the other translated their names such as Sovereign or Quetzal Serpent. I also want to point out the tremendous ambiguity present. One god, Heart of Sky, has three separate aspects or identities, not unlike the Christian holy trinity, which are also supposed to be three aspects of one God.

I find a similar situation with Tepeu and Gucumatz, which means Sovereign and Feathered Serpent; they sometimes seem to be interchangeable with the Framer and the Shaper, and other times they appear to be a separate pair of entities. The same is true for She Who Has Borne Children, and He Who Has Begotten Sons, also translated as the "Forefathers". As I read it, it is not clearly spelled out in the Popol Vuh whether or not all of these different entities aren't just aspects of one couple, a primordial father or male spirit, and a primordial mother or female spirit. That being the case, they may be manifestations of something more like the Taoist concepts of yin and yang, primal female energy and primal male energy.[7]

To confuse the matter more, note that the Feathered Serpent, also known in Mayan texts as the Quetzal Serpent, Gucumatz, or Kukulkan, is the same person known among the Aztecs as Quetzalcoatl, which literally means feathered serpent, or bird/serpent. It's not clear if this was originally a mortal man, a mythic figure, or a god. There are stories that this was a light-skinned man, with a long beard, clad in a long robe, or perhaps in snake scales and bird feathers, that came out of the sea, to help educate and civilize the people; eventually he returned to the sea, long, long before the arrival of the Spanish.[8] But, we know that the name Quetzalcoatl was also sometimes used as the title of a high priest, and a least two actual historical figures bore the name Quetzalcoatl. Perhaps the practice of naming people after a god is not unlike people in Latin America today who are named Jesus.

I have presented here a fairly detailed account of the Popol Vuh to show exactly what was there and what was not. That done, now let me cut through most of this story, to glean a couple pertinent facts related to their myths and prophecies of cataclysms.

1) The Maya clearly did believe that the earth had seen a number of ages of humanity, or early attempts at Creation. And, each time the gods were displeased for one reason or another, and the people (though they be animal, mud, or wood) were driven out or killed, and their world was destroyed. This is a key feature not only for Mayan mythology but, as we shall see, also for Aztec mythology and other mythologies as well. There is some debate whether the story contains 4 previous ages, and we are now in the 5th Age, or whether it contains 3 previous Ages, plus a story about hero twins, and now we are living in the 4th Age. I'm not sure that's too important though. Mythology, after all, is not history or science, and so it needs to be taken for what it is. One thing is clear, the story definitely does contain 5 parts, any way you look at it. Hence we are either at the beginning or the ending of the 5th Age, that part is clear. So, either way it would be auspicious for the Maya priests of old.

2) The second factor is the clear mention of a previous flood. Not only was the world destroyed several times, but the

destruction by flood was one of the most recent and detailed accounts of destruction contained in the Popol Vuh. It was quite specific too. It said that there was a) a great and terrible flood, b) the sky darkened for a long time, c) a great resin fell from the sky, d) it was like a black rain, and e) it rained both day and night. These strange and specific details, more than any other part of the Popol Vuh, suggest that they may not have just been fanciful myths, but an attempt by their ancestors to preserve the story of an epic natural disaster. So, these are key criterion to look for in other mythologies from Mayan literature and from other cultures.

More than that we can't really say at this time. It is a colorful story, but based on this alone, like so many other Mayan scholars, I would not be too concerned about a coming catastrophe, nor would I start building an Ark, just yet.

The Chilam Balam

In addition to the Popol Vuh, there are actually several Mayan books of prophecy. For our purposes I focused on the Chilam Balam of Chumayel, and the Chilam Balam of Tizimin.[9] These are the accounts of two great priests. Chalam Balam means Jaguar Priest, and it is not clear if the title Jaguar means "rich", "important", "powerful" or simply "great" but it is safe to assume that they were high priests. Together they presented a rather impressive collection of prophecies. Some of the most disturbing and alarming prophecies come not from the Popol Vuh, but from the books of the Chilam Balam.

It is interesting to note that some of the prophecies have been eerily accurate in the fairly specific and ancient predictions that they made. They did, of course, predict the return of the Feathered Serpent from the sea. This could very well apply to the image of Cortez the conqueror, in his suit of armor scales with a feathered plume, arriving by ship. The only problem there is that they thought the Feathered Serpent would be their redeemer, not their downfall. Though he may have been a bit of both according to the Mayan books.

In the Chilam Balam of Mani,[10] it predicts that at the conclusion of Katun 13, which is about 1544 AD, the Itza Mayans will see foreigners bringing the sign of the Hunab Ku (One God, or Monotheism). It goes on to describe this "sign of the One God" as an "erect tree". And, it says that this will bring enlightenment to the people. Given the common use of human sacrifice by the Mesoamerican priests, and other practices we might consider primitive or barbaric, prior to the Spanish conquest, this makes some sense. But, even more interesting is the prophecy in Chilam Balam of Chumayel, which adds that there is "no truth in the words of the foreigners."[11] Of course, that is what the Mayan priests said about the neighboring tribes of Itza, when they first migrated to the area.

While the conquest of the Yucatan was prolonged, and there were many expeditions into the Mayan Empire starting in 1525, it was not until 1618 that the first missionaries arrived in the Mayan Itza capital. Nonetheless, the majority of the conquest was between 1530 and 1549. So, the combination of a "foreign people", arriving sometime around "1544", carrying the sign of the "one god", which appears like an "erect tree", and how this will bring some "enlightenment", but that the foreigners are "not" entirely "trustworthy", seems pretty close to 6 out of 6 correct predictions. This is something to think about when considering their prophecies of a future global cataclysm.

The books of Chilam Balam seem to be a rambling list of prophecies. Some are connected with each other, while other verses just seem unrelated. Sometimes, it seems like a long string of unrelated prophecies are assembled in a litany of past occurrences and future predictions.

Sometimes, especially in Chumayel, the author just lists a number of Mayan days, often un-annotated. By the way, when it says something like "6 Ahau" that is a particular day of the Mayan calendar, like when we say September 11th. Here and there, some of the days are annotated with prophecy. The over all format is clearly seen in the following excerpt:

6 Ahau

12 Ahau

10 Ahau

8 Ahau was when Chichen Itza was abandoned. There were thirteen folds of katuns when they established their houses at Chakanputun.

6 Ahau

4 Ahau was when the land was seized by them at Chakanputun.

2 Ahau

13 Ahau

11 Ahau

9 Ahau

(...and so on. You get the idea)

Chalam Balam of Tizimin

Here in no particular order are a number of prophecies from the Chilam Balam of Tizimin.[12] The numbering is my own, for the sake of organizing a long list of predictions, made in these texts. I should first remind you that the end of the Long Count is the conclusion of 13 Bak'tuns, and that date, December 21, 2012, will fall on 4 Ahau on the Tzolk'in calendar. This is usually the way dates are referred to. So, "Katun 6 Ahau" refers to the last day of the Katun, which ends on the day 6 Ahau.

Tizimin Prophecy 1:
> "In the final days of misfortune, in the final <u>days of tying up the bundles of thirteen</u> (bak'tuns) <u>on 4 Ahau, then the end of the world shall come</u> and the katun of our fathers will ascend on high."

For complex reasons involving Itza vs. Classical Mayan timekeeping, we needed to make some adjustments to make sense of this passage. In the original prophecy it says thirteen katuns not thirteen bak'tun, but that would make no sense. The hieroglyphic Bak'tun was probably replaced with the Katun to make it more consistent with the Itza calendar tradition. If you replace the Katun with Bak'tun then you get the real prophecy: "in the final days of tying up the bundles of thirteen bak'tuns on 4 Ahau, then the end of the world shall come". So, once we know how to read it, we see that this is it. This *is* the Mayan Prophecy of Doom that became so famous. And, many believe that the date of "13 Baktuns on 4 Ahau" referred to the date on our calendar of December 21st, 2012.

Tizimin Prophecy 2:

> "The surface of the Earth will be moved. How
> can the people be protected, thus disturbed in
> the midst of the Earth, in the sculptured land of
> Ichcansiho, when all around us there are beggars
> soliciting alms?" … "According to the omens
> above the Earth (stars) and the prophecies, the
> disturbances of our land shall eventually turn
> back, after the years of avarice have passed and our
> sons have used concealment after concealment."

This prophecy alone may not prove so important, it could simply refer to an earthquake, of which there are many in this part of the world. But, I think it is important when combined with other prophecies that we will see. First it seems related to the sad state of affairs on earth, of avarice and concealment. Secondly, it refers to some kind of natural disaster, and not just a quake but something more like a shift, that can then shift back. For now, simply note that it says the earth will move, and then eventually turn back.

Tizimin Prophecy 3:

> "A time will come when the katun-fold will have
> passed away, when they will be found no longer
> because the count of tuns is reunited."

This verse confirms the importance of the beginning and ending of the Long and Short Counts in Maya culture. "The katun-fold will pass, and tuns reuniting", gives us a clue how they perceived time, in general. Time is cyclic and when it ends, it starts over. This was more momentous for the over 5000-year Long Count than it was for the mere 260 year Short Count, but both were a conceptual resetting of the clock.

Tizimin Prophecy 4:

> "The prophecies declared it to the people on that day... As trees grow in the land... our prophecies will prove true. These are the words that must be spoken: The prophecies are a solemn trust from the ancient time. They are the first news of an event, and a valuable warning of things to come."

This verse shows the importance of prophecy in the Maya culture. Interestingly, they see the prophecies as a "solemn trust from the ancient time". This suggests that some of the prophecies may indeed be seen as messages from their ancestors to the present generation, held in trust by the priests for generations.

Tizimin Prophecy 5:

> "Presently, Bak'tun 13 shall come sailing, figuratively speaking, bringing the ornaments of which I have spoken from your ancestors. Then the god will come to visit his little ones. Perhaps 'after death' will be the subject of his discourse. ... My part is to interpret to you. Your part, as well as my own, is to be born again."

This prophecy is the only time that the "Bak'tun 13" is

explicitly mentioned in Tizimin, and it does appear rather ominous. It is odd and important simply because the Bak'tun was just not used in the Itza calendar. This indicates that it is a very special date, unique among Itza calendar prophecies. He clearly predicts that at this time God will come to visit, and to talk to us about the afterlife. Later, he says that all of us must become "born again". This is not typical of his prophecies, yet it does seem consistent with the Tizimin #1 prophecy.

Chilam Balam of Chumayel

There is another book known for prophecies, Chilam Balam of Chumayel, and there is significant overlap between this book and the Chilam Balam of Tizimin. It is as though they are two versions of well-known ancient prophecies. Or, perhaps one priest was an influence on the other. At any rate some of these prophecies are different from Tizimin, and some are especially interesting regarding the foretelling of natural disasters. Here are the prophecies of the Chilam Balam of Chumayel.[13]

Chumayel Prophecy 1:

> "Then creation dawned upon the world. During the creation thirteen infinite series added to seven was the count of the creation of the world. Then a new world dawned for them."

Some 2012 researchers, such as Gerald Benedict, find this an important verse. I think it simply underscores the importance of creation as a cyclical theme, and linked to the number 13, as in the

number of Bak'tuns in the the Long Count. Of course, the more
common and important number for the Maya is the number 20,
as they had a 20 based number system. So, it says "thirteen series
(Bak'tuns) added to seven was the count of the creation". Indicating
that, though they counted in 20s, thirteen was related to Creation.
Stuart underscores this point in his analysis of the end of the Long
Count.[14]

Chumayel Prophecy 2:

> "The katun is established at Uuc-yav-nal in Katun
> 4 Ahau. At the mouth of the well, Uuc-yab-nal,
> it is established…It shall dawn in the south. The
> face of <the lord of the katun> is coverd; his face
> is dead."

This prophecy is unique in that it predicts "it shall dawn in
the south." This could be a fanciful metaphor, but for what we don't
know. It could also refer to a very major earth event, much more
than a mere earthquake, literally the complete movement of the
earth about its axis.

Note that Uuc-yav-nal is the ancient name for the Maya
city of Chichen Itza. This is important because in the next prophecy
it is in Chichen Itza that Gucumatz, also known as Kukulkan, is
supposed to return. And, there is an elaborate pyramid located
there, "The Temple of Kukulkan," which is presumably where
Kukulkan will appear. The lord of the Katun being dead is a
possible reference to the end of time, or the end of the cycle of
time, which is the end of the Long Count, in late 2012.

Finally, the "mouth of the well" is possibly a reference to

the dark rift in the middle of the milky way, sometimes referred to as the womb of the sky. The sun was, in fact, right in the middle of the rift in December of 2012.

Chumayel Prophecy 3:

> "Katun 4 Ahau is the eleventh katun according to the count. The katun is established at Chichen Itza. The settlement of the Itza shall take place <there>. The quetzal shall come, the green bird shall come. Ah Kantenal shall come. Blood-vomit shall come. Kukulkan shall come with them for the second time. <It is> the word of God. The Itza shall come."

This prophecy may be as vague and cryptic as Nostradamus, but there are just too many references to Creation, to ignore. The green bird is the Quetzal. Kukulkan (aka Gucumatz) was present at the beginning of Creation. Now, it talks of the return of Kukulkan to Chichen Itza. This suggests a reference a future creation event, the famed return of Kukulkan. And, with "blood vomit", it sounds like it will be a violent event. One problem here is the reference to the eleventh Katun, but end of the Long Count is Bak'tun 13. So, it is not entirely clear, what they are talking about.

Other Mayan Prophecies

There are a few more prophecies that have been written and talked about a great deal. One is the prophecy on the Tortuguero Stele, and the others are the prophecies of Hunbatz Men and Carlos Barrios, the modern day Maya elders and priests (called "daykeepers"). I think the Tortuguero monument, or Stele, is important because it has actually been at the very center of the discussion on the whole 2012 prophecy, and most all authors refer to it, in one way or anther, either to say how important it is, or to say how it is not important.[15]

The Mayan daykeepers Hunbatz Men, and Carlos Barrios are included because they are not just mindlessly repeating prophecies written long ago; rather they are living, breathing, thinking Mayan priests, who can actually discuss, at great length, what they think the prophecies mean. They also add some prophecies, which are not contained in the 5 existing books of the Maya, but which they feel certain are part of the secret knowledge of the Maya.

The Tortuguero Prophecy:

> The Thirteenth [b'ak'tun] will end
> (on) 4 Ajaw [4 Ahau], the 3rd of Uniiw [3
> K'ank'in].
> Black …[illegible]…will occur.
> (It will be) the descent(?) of Bolon Yokte' K'uh
> to the great (or "red"?)…[illegible]…[16]

Sorry for the poor translation, but this is a highly debatable and yet very important prophecy. For this reason we took every precaution to be as exacting as possible. The reason why this is so important is because it shows the exact date of 13 Bak'tun, 4 Ahau, 3 K'ank'in (December 21, 2012 on our calendar). In fact, there is not much debate about the date on the inscription. It is the overall significance of this monument that is hotly debated. Some say that the date and the reference to gods descending to earth is a confirmation of the 2012 prophecy regarding the end of the Long Count, as is written in the codex. Others say that it is so badly worn that all we really have is a date, and the rest is mere speculation.

The inscription of "Bolon Yokte' K'uh" is clear enough, but what does it mean? It has been translated as "Nine Pole God"[17] or the "nine support gods"[18] The decent of a god to earth would be a momentous event, not unlike the previous prophecy about god coming down to talk to us about the after-life. But who is this god or gods? The nine could refer to the nine gods who rule the underworld, which refers to both death and new creation.[19] At the end of one Sun, or Age, there follows a period of darkness. The nine gods of the underworld corresponds to the nine hells, and they govern the cycle of darkness. This could indicate that December 21, 2012 is the end of our Age, or the end of one time cycle and the beginning of another, as is seen in other dedication monuments.[20]

Tortuguero is one more specific reference to gods coming to earth at the conclusion of the Long Count. And here there can be no doubt about the date. But so much of it is worn away that, except for the date, it's almost unreadable. Nonetheless, there is more than enough to stir controversy and support the idea that the end of the Long Count (December 21, 2012) was in fact an important date to the ancient Maya.

The Prophecy of Hunbatz Men:

> There is a large planet that astronomers do not know about. It is on a very large elliptical orbit, and only comes into the Solar System every 6500 years. When it does, it will have a great effect on

us, in addition to geo-magnetic and gravitational forces acting on the planet. The planet, called Nibiru, an ancient Sumerian name for a mythical planet with a very long orbital cycle, has a number of moons. The 7th moon will activate Kundalini energy in Humanity, actually changing our DNA, to cause higher consciousness and the evolution of the species.[21]

This one is pretty wild. There is an object, according to Men, believed to be larger than Jupiter that is as yet undiscovered by astronomers. It is due to return soon, and it will have a profound effect on human evolution, spiritually, and biologically.

This is very different than our other prophecies of destruction, but it does not preclude natural disasters as well. Who knows what might happen if a planet larger than Jupiter passed closed enough to earth. It would create mass destruction, I'm sure, regardless of whether we grew spiritually from the experience, or not.

The oddest part here is the time-span. One Long Count is about 5,125 years. Five Long Counts are close to 26,000 years. So, where does he get 6,500 years, and how does that fit in with the rest of Mayan cosmology? Well, 4 cycles of 6,500 years are 26,000, same total number of years, but divided into 4 Ages. I don't know how accurate he may be, but this does offer a compelling thesis. Men's prophecy suggests that we are at the end of the 4th Age, and there have been only 3 previous Ages, as it is suggested in the Popol Vuh. True or not, this proves to be one of the most provocative and interesting prophecies so far.

Prophecy of Carlos Barrios:

> Near 2012, the world banking system will start to collapse, and with it our old economic system. The polar ice caps will begin to melt, and sea levels will start to rise. He is optimistic that the world will not end on December 21, 2012, but rather this will mark the beginning of the 5th Age of humanity. In this Age, the world will be transformed, marked by a return of the 5th element, Spirit. This spiritual element will help us to transcend the material world, and create a new way of being for humanity. This will be accompanied by changes in the material world as well.[22]

Like Hunbatz Men, the focus here is not on cataclysmic disasters, but rather spiritual growth. Though, also like Men, he does not rule out physical changes taking place on the planet, but rather he augments those ancient dire predictions with a glow of optimism and the promise a spiritual evolution for humanity. It is interesting that while the predictions of global warming have been made for 20 years, it was not until very recently that the possibility of a banking or economic collapse even seemed possible. So, that's one for Barrios.

Analysis of the Prophecies

I find the prophecies for the end of the Long Count to be few and far between, and very cryptic among the Mayan codices. There are many, many prophecies, but these are the only ones that either stand out for their dire predictions of destruction, or that are related to the end of the Long Count. That being said, there are some specific predictions that are quite unique among the prophecies.

1) There are specific references to the world coming to an end, although some are on 4 Ahau, on what appears to be the end of the Long Count (December 21, 2012), others have different dates, still years ahead in our future.

2) There is a reference to the return of Kukulkan to Chichen Itza on 4 Ahau, which is also a sign of the end of the Long Count, and beginning of a new creation, though there is also a reference to the 11th Katun here.

3) There are multiple references to the earth moving, and even the dawn rising in the South, which would be a major earthquake indeed (if that is even possible).

4) There is a specific reference to both Bak'tun 13, as well as the specific date 4 Ahau 3 K'ank'in, which is December 21, 2012 on our calendar, and it says that then 9 Pole God (or the 9 support gods) will descend, which means that they saw this date as an important event, regardless any other details. Of course, we now know that this date was wrong. As far as I know, no Pole Gods descended on that date. But, if they were off by just 1% of the 5000+ year Long Count, then the prophecy might still hold true for sometime in the next 50 years.

5) At least one modern day Mayan daykeepers states that the end of the Long Count foretells the return of a large, unknown

planet, called Nibiru by the Sumerians. And, there is even greater consensus among daykeepers that the 2012 event will coincide with the biological and spiritual evolution of our species.

Summary

I have to admit, I find many of these prophecies so far-fetched that it makes it very hard to take stock in them. Unknown planets? The dawn rising in the South? Of course, these prophecies are thousands of years old. I think that we can expect some of the messages to be a little garbled over thousands of years of transcription, not to mention perhaps many generations of oral tradition, prior to being written down. But, at least a few of these prophecies are not garbled, they are quite clear. Some of them are simply astounding in their predictions.

At any rate, it doesn't matter. As crazy as these prophecies may be, it's simply our job to see if there is any conceivable credibility to them. I will admit that, at this point, it doesn't look very credible. Perhaps that is why other distinguished Mayanologists have rejected these prophecies. If you just look at the prophecies they look no more credible than any other hokey old myths. But, we are not done yet.

Next we will look at other mythologies and prophecies from around the world. If we can get a match with some of the Mayan mythologies and prophecies presented here, then the Mayan prophecies of doom would have a little more credibility. This is especially so considering how unique some of these predictions are, and how the ancient Mayans were not supposed to have any connection with other ancient cultures from around the world, according to mainstream archaeology.

The goal of this survey of ancient mythology from around the world is to identify a list of myths and prophecies that are corroborated by many if not most cultures from around the ancient world. If we have a short list of prophecies that every major civilization from the past predicted, then we have something substantial that we can subject to scientific investigation.

Otherwise, we are merely engaging in random speculation, based on one anomalous set of prophecies that are unrelated to anything outside of that culture. If no other culture has similar myths or prophecies, then we say these prophecies are a "culture bound" phenomenon, and therefore less likely to be scientifically valid.

3

MYTHS & PROPHECIES OF THE ANCIENT WORLD: AMERICA, AFRICA, SOUTH PACIFIC, AND INDIA.

In the last chapter we focused on the mythology and prophecies of the Mayan texts. But, if their ancient mythology is worth anything at all it would be quite unlikely that they would be the only culture on earth to produce such myths and prophecies. After all, Maya culture was great and advanced but so were the ancient Greeks, and Egyptians, and there have been many other great civilizations from around the world. And what about closer neighbors and relatives to the Maya people, such as the Aztecs, Incas, and Hopi? If the Maya really did record such a major earth event and predict its reoccurrence, shouldn't other myths and prophecies match up with it? In fact, if the myths are accurate then there should have been stories told of such a major event even in the most primitive of cultures; stories told and retold for countless generations of the greatest calamity the world has known.

So, in this chapter we will explore ancient myths and prophecies from around the world to see if there are any correlations to the Mayan myths and prophecies we've researched so far. We start, of course, with their closest neighbors, the Aztecs.

Aztec Mythology

The Aztecs, as mentioned previously, are close cousins to the Maya.

They and the Maya are both Mesoamerican cultures that evolved out of the Olmec culture. They share many features, including the same basic calendar system that each inherited from the Olmecs. They also share common deities such as the Quetzal Serpent, called Quetzalcoatl among the Aztec, which means the same. Like the Maya, the Aztecs also left behind codices and stone monuments with some of their beliefs and mythology.[1] So what was the Aztec Creation mythology, and what if any prophecies did they have about future destruction?

As John Major Jenkins discovered, the Aztec believed that we are living the Fifth Sun, or Fifth Age as we would call it. They believed that there have been four previous Suns, and each one was destroyed. The name of each age is taken from its form of destruction. As David Stuart details in his book and are presented here, the Aztec Suns closely parallel the Mayan previous ages of humanity, but they are by no means identical.[2]

The 5 Suns of the Aztec

> First Sun: The first Sun, named 4 Jaguar, lasted 676 years, and the people of that time were devoured by Jaguars. Then the Sun disappeared.

> Second Sun: The second sun was called 4 Wind, and it lasted 364 years, and was eventually destroyed by Wind, again ending in darkness.

> Third Sun: The third Sun, named 4 Rain, was actually not named after a normal deluge, but rather it was named after a rain of fire, that destroyed the world. This Sun lasted 312 years,

then there was again darkness.

Fourth Sun: The last Sun, 4 Water, was destroyed in a great and terrible deluge, flooding the world, lasting for 52 years, and turning the people into fish. After this, the world was again plunged into darkness.

Fifth Sun: Finally the gods convened to create a new Sun after the previous four had been destroyed. They looked and waited to see from which direction the next Sun would rise. Some looked to the North, others looked elsewhere, while some felt sure that it would rise in the East. Eventually the Sun did rise in the East. The present Age we live in is the fifth Sun, named 4 Earthquake. This world is prophesied to be destroyed by massive earthquakes, and will again bring darkness and hunger.[3]

This is the same essential story as told by the Maya codices, but of course with wildly different details. There are also many interesting similarities. First, you will note that while the time-spans for Suns are brief compared to 13 Bak'tuns, they are nonetheless all divisible by 13!

676 = 13 x 62
364 = 13 x 28
312 = 13 x 24
52 = 13 x 4

The last cycle 52 years is the length of the Calendar Round cycle, of the intermeshing of the 260 day cycle and the 365 day solar cycle, the same as in the Mayan calendar system. So, there is a consistent numerological theme, with both the Aztec Suns and the Mayan Ages, as both evolved from the older Olmec culture.

Naturally there are many differences. The Mayan mythology never talks of any Age called 4 Earthquake or anything like that. Also, the details of how each age ended varies somewhat between the Aztec and Maya myths. The first Mayan Age ended when the animals were condemned to be animals and eaten by each other. In the Aztec myth, the end of the first Sun occurred when all were eaten by Jaguars. Here you see both the similarities and the differences.

In the second Age of the Maya, the mud people were "melted away" and destroyed by their makers but it is not clear exactly how. The Aztec's second Sun was destroyed by wind. Then, the third Age of the Maya, were populated by wooden people, and destroyed by a great flood. The Aztec account is a little different. The 3rd Sun was destroyed by a rain of fire, and then the 4th Sun was destroyed by a flood.

In the Mayan Popol Vuh mythology, it is not really clear which was the third or fourth creation. But the one thing both accounts agree on is that the most recent destruction of the world was that of a devastating flood. In the Mayan text, it appears to be the 3rd of 4 Ages, with an Age of twin-heros inserted between them. Whereas in the Aztec myths the 4th Sun was destroyed in flood, and we are now in the 5th Sun.

This is where the controversy really lies. As David Stuart and others point out, these are clearly not identical accounts of the past. But, as Graham Hancock and others point out, the striking similarities cannot be due to random chance alone.[4] These two myths do indeed read like 2 garbled versions of what was probably a common Olmec mythology, which predated both of them. There are just too many similarities between the two stories. It seems obtuse to dismiss the Aztec mythology simply because each culture embellished the story differently. They both have a story of multiple cycles of humanity flourishing only to be destroyed by the gods. They both have an account of a massive flood as the most recent cataclysm to destroy the world.

Interestingly the 3rd Age of the Maya involves wood people, and the 3rd Aztec sun was destroyed by fire. Obviously a fire is a more practical way to destroy wood people than a flood. Could there be a connection there? There are other similarities as well. The Maya did not name their Ages based on the mode of destruction, but the Mayans did name their Katuns and Bak'tuns for the day that they will end on. So both cultures name the age based on its ending. Again, this is another curious correlation, not identical, but probably more than mere random chance.

Finally, there is one more interesting correlation to Mayan myths. In both accounts there is an issue about the Sun dawning in unexpected places. In the Mayan mythology it said that "It shall dawn in the South". In the Aztec myth it says that at the beginning of the last Sun none of the gods knew from which direction it would dawn. And, of course there are many other references to a new Sun dawning for them, further linking the idea that each Maya epoch of creation was somehow linked to a new Sun that dawned in a new direction. But, it seems completely uncanny that both cultures would link previous cycles of earthly destruction with deviations in the path of the Sun. What flood or earthquake, no matter how massive, could possibly change where the sun rises from? This is another correlation that suggests both mythologies had a common and mysterious origin.

Did the Mayan priests get so involved with mathematics and astronomy that they paid too little attention to these old fireside stories of Creation, and allowed the story to morph into something almost unrecognizable? Like a game played by children, you whisper something into the ear of the child sitting next to you, then they whisper it to the child next to them, and so on. After being relayed 6 times or more, the simple message is usually completely garbled, and often in comical ways. That's what is so fun about the game is to see how mangled a simple message can get after just a few generations. Here, however, it's not so funny. If the ancient Mesoamerican ancestors sought to preserve in myth an account of cyclical global destruction, and a prophecy of future destruction, then it appears that they nearly failed. Each culture distorted that message in their own way, resulting in a badly garbled message. Perhaps we can have better luck with another culture's mythology.

Hopi Mythology

The Hopi provide us with one of the most interesting mythologies of the Americas. They too contain many of the same components in their mythology, plus something else that other mythological traditions do not have. First, it should be noted that the Hopi language is related to the Aztec language, as they are both part of the Uto-Aztecan language group. This indicates that other cultural influences in mythology and ethnicity may also be present in both cultures. So, we would hope to find similarities, but what is interesting is the differences.

The Hopi too believed that there have been 4 previous worlds. They also believed that each previous world had been destroyed. But, they saw it clearly in terms of a spiritual evolution. Each incarnation of humanity would move up to a higher level of spiritual development. Interestingly there seems to be a correlation between these successive levels of spiritual growth and something like the spiritual chakras of the human body.

The first Creation corresponds to the solar plexus chakra, and represents the animal world. The second world corresponds the heart chakra, represents the growth of crafts and villages. The third world corresponds the throat chakra, and represents the growth of cities and technology. The fourth world, corresponds to the mind chakra, and represents the growth of ego, materialism, and imperialism. Finally, the new Age we are soon entering, the fifth world, corresponds to the crown chakra, the most spiritually evolved. This level represents cosmic consciousness. And, there are several other levels of growth to come but, according to the Hopi elders, they are beyond our comprehension.

The Hopi Creation Myth[5]

> The First World: The people of the earth were created in 4 colors: white, yellow, red and black, and they are aware of 5 chakras. The Talker and the Snake make divisions between different color skins. The Creator came and told them that the world will be destroyed and will start over, because

of these divisions and animosity. He said: "follow your crown chakra and you will see a cloud by day and a star by night." This led them to the ant people's kiva (underground hall) and they waited there, while the world was destroyed by fire, because the world at that time was under the leadership of the Fire clan.

The Second World: Already divided by race and separated by nature, the second world went well. The people developed villages. They developed complex arts and crafts. And, life was good, until people began to hoard and barter. They had all they needed but they wanted more. So, this world was also destroyed. This time the hero twins who hold the poles in their place were ordered to leave their posts, and the earth teetered off its axis, and spun crazily, freezing into solid ice. This world had been under the leadership of the Spider clan.

The Third World: Eventually the earth's poles were stabilized and the ice began to melt. This world is known as Kuskurza, for which there is no translation. This world was very successful. The people populated very much. It was populated by big cities with an advanced civilization. They flew through the air on small ships shaped like saucers, called Patuwvotas (literally shaped like shields), and moved so fast they could attack a neighbor and be back before they knew what had hit them. But, the further they went on the road of life, the harder it was for them to hear and praise the Creator. Corruption and war came. All the good people were put in reed boats and sealed up. The earth was then destroyed in a great flood, with waves as big as mountains. When they came to rest, the boat was on an island that had been a mountain top. They sent out birds, one after another, looking for land. Finally the land of the fourth world emerged.

The Fourth World: All of the 3rd world was sunk, all of their proud cities and flying patuwvotas were at the bottom of the sea. The fourth world emerged "Tuwaqachi" (world complete) and the clans were sent out to do their migrations and find the land that each would settle in. Masaw, who was originally an archangel of the Creator in the second world, then became the lord of the underworld and the dead in the third world, was brought back in the fourth world. Now Masaw was to be the custodian of the fourth world. This is our current world. The Hopi people were given a hard land with a harsh environment, but this was good because it kept them humble and respectful of their Creator. Now we are coming to the end of the fourth world.

The Fifth World: There is a song sung at the Hopi Wuwuchim ceremony predicting that the "emergence of the fifth world" would come when the Saquasohuh danced in the plaza. The Saquasohuh is the Blue Star Kachina. Saquasohuh is a far off, invisible star, which will make its appearance soon, marking the beginning of the 5th world. The Hopi elders go on to say that the 5th world has already begun, and that you can see it in the earth itself. Now they are just waiting for the blue star to appear.

In analyzing the Hopi myths, compared to the other mythologies so far, there seems to be a greater degree of human choice in this scenario. Like so many ancient myths talking of people being destroyed for their wickedness, there is a Hopi petroglyph in Arizona, which seems to lay out a clear choice for humanity. It depicts two paths we can take. One path leads to abundant corn, representing abundant life. The other path leads up stairs, representing progress. The path of progress abruptly ends. This seems to say, according to the elders, that the path we take will in part determine our fate, and "progress" is actually a dead-end.

Interestingly, while there is no set time when the new Age is supposed to begin, such as 2012, there are clues. It was said by Hopi elders long ago, that in the end times cobwebs will cover the sky, and that the land will be dug up for treasure. Then, a container may fall from the sky, which will burn the land and boil the oceans. This seems to indicate modern day con-trails, mining, and nuclear weapons. Perhaps this is a marker for the end times, according to Hopi mythology.

Once again we find the same themes as in the Maya and in Aztec myths, and also the theme of spiritual evolution for the species as a whole. Additionally, there are certain patterns that we are starting to see very clearly. All seem to agree that there are at least five worlds; some, like the Hopi, say that we are just now entering the fifth world. Others, like the Aztecs, believe we are at the end the fifth world. The world has been created and destroyed at least three times previously, usually by fire and/or flood. And, now we see what you would think is a unique myth about the earth spinning off its axis, may be more common. The Mayans, Aztecs, and now the Hopi, all describe some deviation in the rotation of the earth and/or Sun. It's curious, one of the most bizarre and unrealistic features of the Mayan mythology appears to be one of the most consistent across cultures, from around the Americas.

What is most unusual here is the mention of a "a far off, invisible star, which will make its appearance soon". This is not consistent with other myths so far, yet at least one Mayan priest has made the exact same prediction. Coincidence? Perhaps. It might be an anomaly, an outlier in the data set. But, it may just be more than that. One thing is for sure, it is a striking detail and stands out among the other myths.

The Inca Prophecy

According to Willaru Huayta, a Peruvian Quechuan elder, in 2013 a giant asteroid three times bigger than Jupiter will pass close to earth. This will activate the "purification" of the land, causing cataclysms and killing most people and animals. But, a core group of people will become the 'seed people' of the next Age.[6]

This is not a prophecy that was recorded centuries ago, and as such it may be a bit suspect. But, with all the different disaster themes in movies and books, it is curious that this one theme is comes up again in different mythologies and prophecies from three different cultures in the Americas, and across millenniums of time. It is present in 3 of the 4 American myths so far. As such, I think we can safely say that we have a new feature to add to our disaster scenario, and look for in other cultures.

The Maori of New Zealand

There is yet another prophecy that predicts a global cataclysm coming soon. Among the Maori of New Zealand, there is a prophecy that around the year 2012 humanity will either ascend to a higher plane or be destroyed, or both. Their ancient mythology predicts that when people become too wicked and sinful, and too preoccupied with their greed and lust, the Rangi (sky) and Papa (Earth) will reunify, violently coming together, and destroy everything. This event, predicted to happen in 2012, is called Ka hinga te aria, which means something like "to go beyond the veil."[7] The veil they speak of is spiritual, and may be synonymous with death. This could mean death to all, or a more optimistic view might mean dissolving the boundaries between the physical and spiritual worlds.

The Maori elders insist that their great grandfathers never heard of the Mayan prophecies, but this old legend of Rangi and Papa goes back as far as anyone can remember. The biggest question here for us is what exactly might this prophecy mean scientifically? In other words, what does the sky and Earth coming together violently mean? Could it be an asteroid crashing into earth? Or, perhaps a cataclysm could occur on Earth either caused by or coinciding with some other heavenly bodies. It is only a myth and a vague one at that, but it is one of a few so far that specifically predict an apocalyptic event around this time in the future.

The Vedic Scriptures

The Vedas are a large body of ancient texts written in Vedic Sanskrit. The texts are oldest in Sanskrit literature, and among the oldest sacred texts in the world. The oldest of them date to at least 3500 years ago.

A key feature of the Vedas are the four Yugas, or Ages of humanity. This is similar to many of the mythologies discussed before, it asserts that there are primarily four Ages, and that an entire cycle of the Yugas takes about 24,000 years, which is close to the length of the cycle of the precession of the equinoxes. The similarity, however, ends there. First of all, they say that the Yugas are all different lengths of time, and all in multiples of 1200 years. So the first Yuga, called Satya is 4800 years, the next Yuga, called Treta is 3600 years, the third Yuga, or Dvarpara, is 2400 years long, and the last Yuga, Kali, is 1200 years long. All total this adds up to 12,000. Then, they believe that the ages reverse themselves and begin going back in reverse order. So, the 24,000 year cycle is really comprised of two 12,000 cycles, one ascending and one descending cycle.[8]

According to the 19th century yogi Swami Sri Yukteswar, in his book *The Holy Science*, written in 1894, he stated that the ascending and descending of the Yugas quite literally corresponds to the ascending and descending of humanity and our physical and spiritual capabilities.[9] Kali is the lowest and Satya is the highest. During the Satya Yuga, humans are virtuous, live much longer, and have special powers. This is the "golden age" of humanity that many talk about in ancient texts, which predates our history. Then with each descending Yuga, people become corrupt, sinful, unhealthy, their height declines, and their lifespan is shortened. The pit of this progression is Kali Yuga. Then we begin going up again.

Some scholars of the Vedas say that each year of the gods is equal to 360 Earth years, so really 1200 years is equal to

432,000 years. This is called the Vedic Long Count. But, Yukteswar disagreed with this and says that, at the beginning of the Kali Yuga in 700 BC, the ancient record keeping became lost, and confused. This was not corrected for two cycles of Kali Yuga, one descending and one ascending. In 1699, however, the beginning of our Age of Enlightenment in the West, we entered the Dvapara Yuga, which will last for 2400 years. So according to according to Yukteswar, 2012 does not appear to be an important date at all. We are 313 years into an Age of humanity that will last another 2,087 years. However he did say something that directly relates to most of the above myths from around the Americas and South Pacific, and he did so very cogently.

Yukteswar gave an explanation for our 24,000 cycle. He said that our Sun "takes some star for its dual, and revolves around it".[10] This guru from 19th Century India came up with a very solid scientific thesis that the entire solar system is actually part of binary star system, with an elliptical orbit of 24,000 years. We spend 12,000 years moving away from it, and 12,000 moving toward it. We are now moving toward it, and picking up speed as we do. He says that this explains the precessional cycle, which he thought was 24,000 years long. But it places our next rendezvous with our binary companion in about 10,000 years from now, not anywhere near our time.

He goes on to say that it is when the Sun and our planets are closest to our Binary companion, that people will become so developed mentally, emotionally, and spiritually, that we will be able to comprehend all. In this scenario there is no cataclysm, it's all good for humanity. But, if such a thing did happen would it not cause some disturbance on earth, maybe earthquakes or tidal waves?

Once again, we have another ancient mythology, the details written out and commented on over 100 years ago, completely independant of the myths of the Americas, that predict a large heavenly body coming close enough to the earth to make dramatic changes, and usher in a new era. The difference is that the timing and the cause is quite specific and quite different from the Mayan, Aztec, Hopi or other myths.

The Dogon Mythology of Mali, West Africa

Finally we turn to the tribal myths of Africa. The Dogon, a tribe in Mali, apparently have known for centuries that Sirius was a binary system, and that Sirius B is a white dwarf. In 1976 they also said that there was a third star, a red dwarf. Later in 1995, astronomers Daniel Benest and J. L. Duvent published, in the Journal of Astronomy and Astrophysics, the discovery of Sirius C, a small red dwarf, just as the Dogon predicted. They said that this information was given to them by visitors from the Sirius system. "They will return someday, and it will look like a new star, and will become visible in the night sky."[11]

So here is yet another myth, one that is apparently not related in any way, except in that it also involves the appearance of a new star in the sky. And, more important, they have obscure scientific information that they should not have, and yet they do. If these myths are indeed garbled messages from our ancestors, then it appears that the Dogon myths are a bit less garbled than others, albeit incomplete. And, this now adds to anther prediction of an unknown heavenly body making its appearance in the night sky, as a sign of the coming age.

The Zulu Prophecy of Mu-sho-sho-no-no

The Zulu of South Africa believe that there is an unknown star
called Mu-sho-sho-no-no, which sometimes orbits close enough to
the earth to be visible. They say that this star is made of the lava of
our Sun. The last time this star became visible, it "turned the earth
upside down" so that the "Sun rose in the South and set in the
North". Then came "drops of burning black stuff" after a terrible
deluge with great winds. This, we are told, is going to happen again
in the year of the Red Bull. Interestingly, it appears that the year of
the Red Bull came again in 2012.[12]

    This is clearly one of the most crucial prophecies of all,
because it weaves several different unique details together, in a way
that would seem to defy mere random chance. Again, we have this
unusual account of the Sun rising in an unexpected location. Also,
the Mayan mythology talked of a "black rain" and a "heavy resin"
that came down from the sky. Then the Aztecs talked of the world
being destroyed by a rain of fire. Now we hear of a very similar
report of "drops of burning black stuff".

    Finally, there is again the warning sign of a new star in the
sky, but this time it is part of a more cogent message. This time
the star seems to "turn the earth upside down". This is something
more like a causal relationship between this warning star, and the
now familiar phenomena of the Sun rising in a different direction.
It seems to make sense that a large massive object like a star, could
actually pull the earth off its axis, if it came close enough.

    This is clearly an important myth in that it acts as a
nexus for features from different myths we've looked at so far, and
combining them in a more cogent manner. But what of Sirius? If

we put this together with our other mythologies from Africa, the Americas, and India, it suggests that the Sirius star system is on a long orbit with our Sun, 24,000 years, and when it returns it causes a great and terrible cataclysm on Earth, and ushers in a new and improved age of humanity afterward. And, humanity has recorded at least 3 or 4 of these orbits so far.

Is this possible? It seems far more fantastic than the Mayan myths we started with, which were fantastic in their own right. Perhaps it is nothing more than coicidence. After all if you took any 6 or 7 myths from different cultures you are bound to come up with some coincidneces, and then you could weave a common myth, which includes elements of each. So, lets look at another culture, one that is a bit closer to home, culturally speaking, Ancient Greece.

4

MYTHS & PROPHECIES OF
ANCIENT GREECE

I go next to Plato for the simple reason that Greek mythology is so
much better known and understood in Western culture. Here we
are not dealing with some alien set of concepts, numbering systems,
calendars, and gods, but rather with the very foundation of Western
civilization. So, what, if anything, does Plato and Greek Mythology
have to say about our doomsday scenario? You'd be surprised.

Plato is famous for his philosophy and his dialogues with
Socrates, but he is also famous for his ancient myth of Atlantis. It
was in his dialogues of Timaeus and Critias that he documented the
now famous legend of the lost civilization of Atlantis, which was
destroyed in earthquakes and floods, and "in the course of a single,
terrible day and night" sank beneath the sea and vanished. The
legend may have been known by others of his time, but it is only
in his writing that we have any documentation of Atlantis from the
classical period.

This one myth is really the archetype for much of the
antediluvian mythology and speculation in Western civilization,
because it has all the basic elements. It is an undiscovered
civilization that existed long before recorded history, about 11,500
years ago according to Plato, and it was an extremely advanced
civilization, by classical standards. Plato reports that they "drew
their water from two springs (one of cold and the other of warm

water), ...which it produced in generous quantities." Clearly, it sounds like someone describing hot and cold running water, using the vocabulary of classical Greece.

Regarding the Mayan prophecies, one of the most important parts of the story is his description of a cataclysmic flood combined with massive earthquakes that destroyed both Athens and Atlantis, completely sinking Atlantis to the bottom of the sea.

There are several elements to Plato's story that make it compelling. First, he doesn't report this as a Greek myth, and he doesn't credit it to an anonymous source, but rather he provides a provenance for the story that leads us to ancient Egypt. He is allegedly repeating a story told by Solon, a relative of Critias's great-grandfather who visited Egypt some 100 years earlier, and there met with a very elderly Egyptian priest. The priest, according to Plutarch (46-120 AD), was named Sonchis of Sais, and he chastised Solon and the Greek culture for their lack of historical knowledge.[1]

> "You have no ancient tradition to imbue your minds
> with old beliefs and with understanding aged by time.
> The reason for this is that the human race has often been
> destroyed in various ways – as it will be in the future too.
> ...You remember just the one deluge, when there
> were many before it. ...but none of you realizes it, because
> for many generations the survivors died without leaving a
> written record. But in fact there was a time, Solon, before
> the greatest and most destructive flood..."[2]
> -Plato / Timaeus

Now it is easy to see why I was so lenient in accepting minor variations between the Aztec myths and the Mayan mythology. This is a classic case of a mythological theme and variations. Here, Plato basically tells the same story again, probably written about the same time as the end of the Olmec civilization (4th to 5th Century BC), but with none of the embellishments of the Mesoamerican mythology. Again, the theme is of reoccurring major cataclysms, destroying previous civilizations, and nearly killing off all of humanity. He says it has happened a number of times, in different ways, mostly from fire and water, and that one of the most recent events was that of a great flood. All of these details correlate with the American myths.

Finally, it is important to note that he states quite clearly that it will happen again. This is not so much in the form of a prophecy, but rather an almost scientific certainty. How does the Egyptian priest from the 6th Century BC know it will happen again? Plato does not say.

Sonchis goes on to explain that the Egyptian culture is much older than the Greek civilization, and they have more of an unbroken record of history, aided by the early invention of writing. "This explains why the legends preserved here (Egypt) are the most ancient, even though the human race is actually continuous, in larger or smaller numbers, everywhere in the world." Sonchis goes on to say the following:

> "After the usual interval, a heavenly flood pours
> down on you like a plague and leaves only those
> who are illiterate and uncivilized. As a result,
> you start all over again and regain your childlike
> state of ignorance about things that happened
> in ancient times both here in your part of the
> world."³
>   -Plato / Timaeus.

After the "usual interval"? That is an extremely interesting turn of phrase for Plato to use, in that it clearly reinforces the idea that the Egyptians were in fact certain that there would be another such global destruction, after the "usual interval" of time. And, they might very well have known what that interval was, and when to expect the next major cataclysm. If they did, Plato did not record that bit of information for us.

How, you might ask, do we know that this was a global cataclysm? After all, many have suggested, and indeed the archaeological record indicates, that the lost civilization of Atlantis was probably an old myth based on the actual destruction of the Minoan civilization. We know for a fact that the Minoans were destroyed a thousand years before Plato, by a combination of a devastating volcanic eruption, earthquakes, and most likely a 100 foot high tidal wave causing a major tsunami. And we now know that they were an advanced civilization with hot and cold running water, indoor toilets, and many other features of modernity. But, there are other reasons to believe that Plato was referring to another, larger, global cataclysm.

In Plato's Timaeus dialogue, Sonchis goes on to retell the myth of Phaethon, scion of the Sun, who "harnessed his father's chariot, but was incapable of driving it along the path that his father took, and so burned up everything on the earth."[4] This is the myth of how the son of the Sun in the Sky tried to drive his "father's chariot", in other words, to steer the Sun through the sky, but could not control it, and burnt the earth. But, Sonchis informs Solon where this myth originally came from.

"This story has the form of a fable, but it alludes to
a real event – the deviation of the heavenly bodies
that orbit the earth and the periodic destruction at
long intervals of the surface of the earth by massive
conflagrations."[5]

-Plato / Timaeus.

This would be interesting enough as it is, but it is even
more interesting when combined with another reference in the
Statesman, by Plato, in which he refers to the myth of Atreus and
Thyestes.

"There really did happen, and will again happen,
like many other events of which ancient tradition
has preserved the record, ...the story, which tells
how the Sun and the stars once rose in the West,
and set in the East, and that the God reversed
their motion."

...There is a time when God himself
guides and helps to roll the world in its course;
and there is a time, on the completion of a certain
cycle, when he lets go, and the world ...turns
about and by an inherent necessity revolves in the
opposite direction.

...And the world turning around with
a sudden shock, being impelled in an opposite
direction from beginning to end, was shaken
by a mighty earthquake, which wrought a new
destruction of all manner of animals."[6]

-Plato / The Statesman

This passage is unique in its detailed description of a global calamity and his insistence that this old myth is based on ancient, long forgotten, actual events. Most important for us, it brings up a myth about a natural disaster that involves the Sun rising in the wrong direction, caused by a heavenly object with a long interval (orbit) with earth. How many cultures now have said the same thing? The Maya, Aztec, Hopi, and now the Greek and Egyptian mythologies, all include a deluge, possible earthquakes, but most bizarre of all, every one of them includes a reference to the Sun rising in the wrong or an unexpected direction.

This truly does seem to be beyond random chance. It is like a tell-tale sign woven throughout different myths, that they are all referring to some very real, very specific, common, global event, garbled and filtered through thousands of years, yet still quite specific. And even more bizarre is that the Hopi, Inca, Vedic, Dogon, Zulu, and now the Greek and Egyptian myths all speak of a rendezvous with a heavenly body on a long orbit around our sun, possibly a star, which coincides with or causes these cataclysms to occur.

To be fair and accurate, I feel the need to mention that Plato's account is still a form of Greek mythology. Though we often put Plato and Greek mythology in different categories, it is important to note that, in the above passages, Plato is retelling stories and myths, and not deducing truth from rational Socratic tradition. Even if Plato's own great grandfather was the fabled Solon, who met with the Priest Sonchis, the priest's tale is nonetheless hearsay. And, there is no continent such as Atlantis where Plato said it was, beyond the pillars of Hercules (now known as the Strait of Gibraltar). None, that is, except for the Americas. Of course, this only means that the writings of Plato are on the same level as the Mayan Popol Vuh or Chilam Balam. In other words, here we have the mythical writings of someone who in Western civilization is as esteemed and revered as the ancient Chilam Balam was in the Mayan world. Though he may have merely repeated myths of Greek culture, he clearly believed that these particular myths were based in fact, albeit long forgotten. So, Plato, in a sense, vetted the Greek mythology that he thought was most valid and relevant to us.

What is really most amazing about Plato is that he seems to speak our language. He takes the same themes, the same mythological history and prophecies of the Mesoamericans, and words them in a way that sound more rational to someone from Western culture. Plato is, in a very real sense, a missing link between the bizarre myths and prophecies of the Maya Jaguar Priests and our modern, scientific world.

Plato cites very similar myths as in other myths, and then says to us quite pointedly that some of these myths are based in fact, though long forgotten. And, he has a priest from what is arguably the oldest civilization on earth (Egypt) to support his claims. This seems to carry more weight than just myths or fables. And, most interesting, these stories are quite distinct from other Greek and Egyptian myths. Yet, it does explain how the Greek and Egyptian myths developed, by giving us a more cogent account.

5

MYTHS & PROPHECIES
OF ANCIENT MESOPOTAMIA:
NIBIRU, GILGAMESH, AND NOAH

The ancient Sumerians have a legend of a mysterious star called Nibiru, which means 'crossing'. This star has been alleged by modern writers to be an unknown planet, or perhaps an unknown binary star in orbit with our own Sun. Others say that they were referring to Jupiter, but they had another name for Jupiter as you will see. So, the myth is included here for the sake of completeness, but the details of this myth are vague at best. Below you see the myth of Nibiru, as taken from different passages mentioning Nibiru in ancient Sumerian cuneiform tablets.

> Nibiru, which is said to have occupied the passageways of heaven and earth, because everyone above and below asks Nibiru if they cannot find the passage. Nibiru is Marduk's star which the gods in heaven caused to be visible. Nibiru stands as a post at the turning point. The others say of Nibiru the post: "The one who crosses the middle of the sea (Tiamat) without calm, may his name be Nibiru, for he takes up the center of it". The path of the stars of the sky should be kept unchanged.[1]

There are several other vague mentions of the star Nibiru in Sumerian texts. In one it even lists Marduk, Nibiru, and Jupiter, as though they are all synonyms for each other, but in the same passage it is described as a big star though its light is dim, which does not sound like Jupiter. From the above quotations I think you see that there is not a lot to go on in the form of a specific prophecy or mythological story. Suffice it to say that the ancient Sumerians, one of the oldest civilizations on earth, believed that there was a very big, but very dim, heavenly body that "crossed" the sky, and that this was obviously an important omen for them and their people. It is in fact referred to as a "post at the turning point," presumably like a signpost.

Nibiru means 'crossing point' or 'ferry' and we see this reference in the above passage. The idea of crossing could, of course, refer to a dramatic motion across the sky, such as a comet or the movement of a planet. It could also refer to the end of an elliptical orbit, in which a star or planet comes quickly and close to our solar system, before moving back out into space. But, more than that we cannot say.

It is interesting to note that Marduk was the calf of the Sun god Utu. In other words, Marduk was the son of our Sun god. Suggesting perhaps that Nibiru was the offspring of the Sun. Here we see a connection to the Greek myth of Phaeton, the Sun's heir who tried unsuccessfully to steer the Sun and ended up burning the earth. And, we also see the idea that there is a big Sun and a little Sun, and Nibiru is the little Sun. In this way, it seems clear that they are describing a binary star system, with our Sun being the larger and brighter of the two, and with a second, dimmer star orbiting it.

Note that the above passage ends with an interesting statement, "The path of the stars of the sky should be kept unchanged." Why would they change? The path of the stars is the same as the path of the Sun; if one changes path, presumably so too would the other. So, we see the same reference to the Sun and stars changing their path in the sky again, as in other myths, this time written a little differently. This time it is almost a plea for the stars not to change their path again.

This now forms a deep, widespread, and fairly consistent pattern of mythology from a variety of sources all referring to a heavenly body, perhaps a large planet, or a dim star, which will become visible when close enough to earth. They say that it is like a signpost signaling some critical juncture in time. And, they too believe that it is related in some way to changing the path of the stars and Sun. This, combined with other mythologies, forms a fairly consistent pattern seen in so many ancient myths, across different cultures and across eons of time. But, the Sumerians added a companion detail, that it is named the "crossing star" as it is known to "cross the sea." We can only guess what sea a star would cross. Perhaps the sea they speak of is the dark rift in the Milky Way, the sea in the middle of the starry sky, like the 'well' in the sky, mentioned in Mayan prophecy.

The Deluge of The Epic of Gilgamesh

Perhaps the granddaddy of all flood myths is Noah's Ark. At least to most, this is the most familiar tale of a cataclysmic deluge in which humanity was wiped out, and only a small group of people survived. We will go through the Biblical version of this story blow by blow, but first it should be mentioned that this story was probably based on another Sumerian myth, the Epic of Gilgamesh. We already covered the Sumerian myth of Nibiru, but now we have this tale of a great flood as well. This myth, combined with the Nibiru myth, form many of the features of other myths we've seen from around the world. This story, which is probably an older version of the Noah story, is actually a story within a story. In one section of the Epic of Gilgamesh there is a story of a man named Utnapishtim. This is his story.

The Story of Utnapishtim (from the Epic of Gilgamesh)

> Utnapishtim tells Gilgamesh about the story of his life. How the gods conspired to destroy the world, and not tell anyone. One god, Ea, decided to warn Utnapishtim. The god commanded him to build a large boat. Utnapishtim obeyed and built a huge ark, 120 cubits long (200 feet). 10,800 units of bitumen (organic tar) were used to seal the hull. And, 10,800 units of oil were stored on the ship. All their animals, family, relatives, craftsmen, and all the beasts and animals of the field were put on the ship.
>
> Once they entered the ship, they sealed the entrance, so the ship was completely weatherproof. A huge and dark cloud formed on the horizon. The storm was like nothing else in history. The gods worked together to attack the earth. There was complete blackness, a wall of

water engulfed and overwhelmed the people. The
land was shattered like a pot. The people filled the
sea like fish. The storm lasted 6 days and 7 nights,
and on the seventh day it pounded like a woman
in labor.

After the storm, all was calm. The land
had been flattened, and all the people turned to
mud/clay. Utnapishtim opened a window breathed
the air, and fell to his knees in prayer. The boat
eventually came to rest at the top of Mount
Nimush. He sent out a series of birds to see if
they would return. First a dove, then a swallow,
then a raven. When the raven did not return, he
let loose the animals.

Upon the completion of their ordeal,
Utnapishtim made a sacrifice to the gods. The
great goddess was pleased with Utnapishtim but
she was angry with the god Enlil, whose idea it
was to destroy humanity. In exchange for his good
work in saving what was left of humanity and the
animals, Utnapishtim and his wife were granted
immortality.[2]

What is so special about this story is not that it is an exact match
with other mythologies. It certainly has none of the unique features
listed before in the other myths. But what is important about this
story is just how wide-spread this narrative is.

If you are familiar with the story of Noah's Ark, you
recognize the details instantly. This is probably the most common
of all flood myths. Just look at the story of Noah, for comparison.

Point by point, the stories parallel each other almost exactly.

Noah's Ark (Genesis 6-10)

Because of humanity's wickedness, one
day God decided to wipe them out, to "blot out
from the face of the earth all mankind." But, one
man Noah, was a pleasure to the Lord. So, God
told Noah of his plan to destroy the world, and
warned Noah to build a boat, actually a great three
story ark, 300 cubits long, and 50 cubits wide
(450 feet long and 75 feet wide), and 30 cubits
(45 feet) high. He was told to cover the entire
boat in pitch (tar) and place inside the boat a pair
of every animal, bird, and reptile, both male and
female, and keep them alive during the flood.
Then he was told to store in the boat all the food
that he and his family and the animals would
need. Noah did as he was commanded.

One day the Lord told Noah it was time
to board the boat. One week later the Lord began
the deluge. Noah was 600 years, 2 months, and
17 days old at this point. For 40 days and 40
nights the roaring floods prevailed, covering the
ground and lifting the ship high above the earth.
The sea rose 22 feet above the highest mountains.
All existence on earth was blotted out. All living
things that breathed upon dry land perished
except Noah, his family, and his ark of animals.
Finally, 150 days after the flood began,

the ark came to rest on top of Mount Ararat. After 40 days Noah opened a window and released a raven, which flew back and forth until the earth was dry. He then released a dove, which quickly came back. The next week he released the dove again, until finally the dove returned with an olive branch in her beak, showing Noah the waters had gone down. 29 days later, Noah opened the door and released the animals, birds, and reptiles.

Noah then built an altar and made a sacrifice to God. Jehovah was pleased with the sacrifice and promised never to do it again. God, then blessed Noah and promised him and his family a long and happy life, and made him master of all the lands.[3]

As you can see from this, the stories of Noah and Utnapishtim are extremely similar. But what is really amazing about these stories is that it is not just ancient Sumerians and Hebrews that had such stories. In fact in Graham Hancock's book, he presents 25 separate ancient flood myths from around the world, all with an almost identical detailed narrative.[4] He presents 3 similar myths from Mesoamerica, 6 from South America, 8 from North America, 7 from Asia, 1 from India, and 4 from Egypt, Greece and the middle east. He goes on to state that there are over 90 such stories from cultures all over the world. Of course they vary a bit from tale to tale, and few are as similar to each other as Noah and Gilgamesh. But after reading all 25 of the above myths, most include at least some of the unique details of the Gilgamesh narrative, such as being warned by a god, building an arc or some sealed craft, often collecting birds and animals to save, being sealed up in the craft and riding out the storm. Then, landing on a mountain top and releasing birds to see if the waters had receded and finally leaving the arc and populating the earth. In fact, according to Hancock, this myth is even more pervasive than the myth of Quetzalcoatl, which he says is also a common mythological narrative found in South America, Mesoamerica, North America, Egypt, and ancient Sumer. In fact, the story of Quetzalcoatl seems to be a companion story to the flood myth. It takes place after the flood, when members of a civilized race roamed the earth by boat, helping the survivors.

The Myth of "Civilizers" from the Sea.

> After the last great deluge, a white man with a long robe and a long beard came out of the sea. He brought with him advanced technology, philosophy, morality, and helped the survivors rebuild civilization. In South America he his known as Viracocha, while in Mayan myths he is known as Gucumatz, or Kukulkan. Among the Aztecs he his known as Quetzalcoatl, all these names mean the same, feathered serpent, because the stranger had strange scales on his robe, and wore feathers on his head. Among the Hopi, he is known as Pahana, the "true white brother." In ancient Egypt, a parallel tale is told, though greatly embellished and mythologized, about the god

Osiris.

Most of these myths also refer to his constant companion and helper, who was named or depicted as a dog.

In each case, these great civilizers taught the people much of what we consider advanced civilization, including the promotion of peace, morality, and the teaching of technology and agriculture. But in most myths the wicked people rose up against the stranger and his helpers. The pale strangers were either killed or driven back into the sea, depending on the legend. Most agree that he will return. The Mayans expect Kukulkan to return to Chichen Itza at the end of this Age. And, the Hopi predict that Pahana will return when the Blue Star Kachina removes his mask (appears in the sky for all to see).[5]

Summary: Lost Knowledge vs. Old Stories

From all these myths and stories from around the world, presented in Part I of this book, it seems to me that two different types of mythologies have been preserved for us. One, such as the Mayans, is more prophetic and has a more sweeping Creation narrative. The other, such as Noah, is more specific, more recent, and more regional in nature.

If you compare and contrast the Mayan prophecies with the stories of Noah, you see a stark contrast. Noah, Gilgamesh, and even Quetzalcoatl, and similar myths are more numerous in quantity around the world, yet lack important details about what caused past cataclysms, and lack specific future predictions of cataclysms. It is as though these flood and reconstruction stories were a very common scenario, which many different cultures experienced in the last deluge.

I don't know why these tales are so common, and why these tales often do not contain the other more prophetic details mentioned before, but in general the Noah narrative, and even that of Quetzalcoatl, seem to be a more common myth. And, I don't just mean common as in numerous or frequent. I mean that it appears to be a tale that could have been repeated among the common folk from all over the world. It is the simple story of how they survived the last great deluge, and the strangers who came to help them rebuild. By contrast, the Mayans, Aztecs, Hopi, Egyptians, Plato, and the Vedic Scriptures seemed to have also had a secret knowledge that could not possibly be known by the average person, and certainly not refugees from a relatively backwards settlement.

It is logical to assume that some cultures had a more robust and well informed priesthood or academic class than others. The more advanced civilizations perhaps attempted to retain their secret knowledge through elaborate mythology and prophecies. Meanwhile, the common people simply told and retold the how their ancestors "survived the great flood" without knowing how many times it had happened before, what caused it, or how it was related to the stars or the path of the Sun.

If these myths are garbled messages from the past, then the above mythologies constitute two different garbled messages. You might say one story is passed down from the priest/philosopher's child, and one is from the farmer's child. For our purposes we are concerned with that of the priest's tale. This is the type of tale we see in the Mayan prophecies.

Secret Knowledge of the Ancients

The Mayan prophecies of doom, though not appearing to be a focus of Mayan culture, nonetheless reverberate and resonate with other ancient mythologies from around the world. In fact, one could argue that the Mayans are not even the best or most complete of these mythologies. What started out as a quest for answers among the Mayan texts has revealed an astonishing fact, that almost every ancient major mythological tradition from around the world has the same common features. Furthermore, some of these common features are so bizarre, and one would think unique, that it is hard to believe that it is just a coincidence that they pop up in so many separate ancient mythologies.

Floods are common enough, but black burning resin? That sounds like a real phenomenon, perhaps the ash from a volcano, mixed with rain or other condensing gases? And, what about the Sun rising in another direction? That seems completely bizarre and unrealistic, and yet it is a key feature in all of these myths. Finally, what about the "heavenly body" or unknown star, which will appear? The image or sign of a star is common enough, such as the star of Bethlehem, but the way it is discussed in these myths seems more scientifically cogent and specific than that. It is not just the sign of a star that will appear as an omen, but it is often referred to specifically as an unknown, dim, planet or star, which orbits our Sun at long intervals. This seems to be different than something you'd expect in most of the campfire stories of our caveman ancestors.

The Mayan prophecies, as well as that of the Aztecs, the Hopi, Egyptians, the Vedic Scriptures, and even the writings of Plato, and some other myths such as that of the Inca and Zulu tribes, seem to grasp the big picture. They look at all of humanity and all of history, and attempt to contextualize it in some rational structure. They provide the most cogent explanation of what did happen in prehistory and what will happen again in the future. This then should constitute our list of mythological details, which we will research in the next section. These are specific features of the Mayan mythology and prophecies, which have been compared to and vetted by other corroborating mythologies from around the world. All of these myths contain certain common elements. We summarize all these common details in the following overall mythological scenario.

A Composite World-wide Prophecy of Global Cataclysm

> a) There have been multiple cycles of world destruction, with the number five being the most common; we are either at the start or end of the fifth Age of humanity. And, we are doomed to repeat this cycle of destruction again and again.

> b) Usually the destruction involves fire or flood or both, but two other features are common as well, a burning black rain/resin will fall, and the earth will fall into darkness.

> c) Almost everyone of these mythologies report a shifting of the earth's axis, or some other

phenomena involving the Sun rising in a new and different direction.

d) Many, if not most, myths report that there is a large heavenly body, either a star, a planet, or an extremely large asteroid, that only returns to earth after very long intervals. It will be seen as a new star in the night sky when close enough. It will either signal the approach of, or actually cause, the above apocalyptic scenario.

e) This cataclysmic event will either signal the coming of, or directly cause, some advance in human evolution or positive change for humanity, for those who survive.

One feature seems lacking from the above list, and it is perhaps the most important. Most of these myths do not predict any specific time frame for a cataclysm to occur. There are exceptions, of course. The Maya especially, and also the Maori and the Zulus each specifically predicted 2012 as the year of a great change. One Inca elder said it would happen in 2013. The Hopi and others just indicate that it will happen soon. And, the Vedics are quite specific that it will not occur for another 10,000 years.

Edgar Cayce's Predictions

One place to look for confirmation of a timeline for a reoccurring global cataclysm is the 20th Century psychic Edgar Cayce. This might seem an unlikely source, but the "readings" he gave, while unconscious and deep in trance, were amazingly accurate. He could even accurately diagnose illnesses in people, and did so repeatedly. In some of his readings he specifically referred to the destruction of Atlantis and predicted that such a global cataclysm will occur again in the near future, and he gave a timeline for these events.

Cayce reported that the destruction of Atlantis occurred around 11,500 years ago, just as Plato had reported. Of course, Cayce might well have read Plato, so this is not so incredible. What is more interesting is his predictions for the future. He detailed specific events of a future global cataclysm, such as major earthquakes in California, title waves, and rising sea levels, which will flood much of the Western United States. He said this would be a global event affecting all the continents on earth. And, he even gave us a map of what the earth would look like in the future.

Amazingly, he predicted that there would be a physical shifting of the earth's poles, causing the ice caps of the Arctic and Antarctica to melt, and causing other areas to freeze and become the new polar ice-caps. This certainly seems to fit with other myths and prophecies we've seen from around the world. And when did he predict this would happen? He did not predict a specific year, but he said that it would occur with the new millennium (sometime around or after the year 2000). As vague as that is, it does generally fit with other ancient prophecies we've seen.

We will need to take this into consideration, as we move to the next section on scientific theories. Our goal now is to take the specific features of this doomsday scenario, and to research them against established scientific theories, to see if such wild and fantastic events are even possible, and if so, how. Could it be possible the ancient myths were observations of a real scientific phenomenon, preserved in the form of a garbled message from the past? Let's see.

Part II

SCIENTIFIC THEORIES OF GLOBAL DESTRUCTION

6

PRECESSION OF THE EQUINOXES

In this section we will look at scientific theories of how global cataclysms might have occurred in the past, and may occur again. We will focus only on those theories which are capable of explaining the specific details listed in our mythological scenario, presented at the end of the last chapter. If a theory does not explain how the Sun might rise in the wrong direction, or how that might be related to a new star in the sky, then it doesn't really help us. First I think we need to explain the scientific concept of precession because it comes up a lot in various theories of global disasters.

What exactly are we talking about when we refer to precession? What the heck is that anyway? I'm sure many reading this have had that thought. But I think it's important to separate out mythology from real science. So I saved precession for this section, on the science of global destruction. Precession is based on real science. As we will see, there may be some scientific debates about the details of precession or the implications of precession, but there is no debate about the existence of precession. It is a scientific fact. What follows is a brief description of what precession is, and at least a couple theories of what causes it.

Have you ever seen a top spinning, or a gyroscope? They not only spin around on their axis, but the axis also spins around in a circle, albeit much slower than the spinning of the top. Especially with a gyroscope, the axis is quite prominent, and you can often see that it is not pointing straight up and down, but usually pointing off-kilter. Slowly that point moves in a circular motion, so that the top pole of the axis carves out a little circle. It points north, then northeast, then east, then southeast, then south, and so on, in a circle. That, in a nutshell, is precession. As a verb, you might say that the gyroscope precesses around in a circle, or that it is precessing.

We all know that the Earth is like a giant top, spinning around its axis, as it orbits the Sun. And, we know that the Earth's axis is also off-kilter by about 23°. So it should not be surprising that the Earth also precesses about its axis. The biggest difference is the speed. A top spins very quickly, making many revolutions every second. The earth spins very slowly, taking an entire day just to make one revolution. So, if it takes a top a full 10-20 seconds to precess around in a circle, then it should not be surprising that it takes the earth many thousands of years to complete one circle of precession, approximately 26,000 years. Actually, that number has recently been revised, but we'll get back to that later. The ancients referred to this cycle as "The Great Year."[1]

Precession of the Equinoxes

So, how does one monitor or perceive precession? Well, if we take the same date every year, let's say the spring equinox, and we look at the stars relative to a fixed point on the horizon, we will see that they are moving very slightly from year to year. Not a lot, and not quickly, but definitely moving during the course of a lifetime, about 1° every 72 years. So if ancient astronomers had an observatory, such as the ancient Mayan observatory at Palenque or Stonehenge in England, then they would have fixed markers to compare the star's positions to, from year to year. They could monitor the stars position on certain dates, relative to the fixed markers. Then, if they had written records as the Maya did, they would note that after several centuries the stars were not lining up with those fixed markers, as they did in their great-grandfathers time.

This is a very well known phenomenon in Western culture. Most people are familiar with astrology and the zodiac. And they've probably heard of the New Age or the Age of Aquarius. Well, that's based on precession. You see, all the constellations of the zodiac lie on a path in the sky. That is the path that the Sun travels in the sky every year, as we orbit the Sun. So, every spring equinox the Sun appears to be in front of the same constellation of stars. Well, that shifts about one constellation every 2000 years. That is why they say that we are entering the Age of Aquarius. It is because this year, at a certain point, the Sun will appear to be edging into the constellation Aquarius. For the last 2000 years the Sun appeared to be in front of the constellation Pisces, on that same date every year. In another 2000 years it will be entering the constellation Capricorn. And, so on. You will notice that precession moves backwards, the opposite of the normal Zodiac, in which Pisces follows Aquarius, which follows Capricorn.

Note that astrology is not the same as astronomy, and there are many differences between the two. But for our purposes it helps to give you an idea of what Precession is all about. Over the course of thousands of years, the Sun does appear to be precessing through the constellations of the zodiac, only backwards.

This leads us to the question of what did the Maya know, and when did they know it? Did the ancient Maya astronomers observe the precession of the equinoxes as described above? Well, they did observe the heavens very carefully, and they did take detailed written notes, and their civilization did last for many centuries, so conventional wisdom would say yes, they probably did. This topic, however, is not without argument. Specifically, certain archaeologists, such as David Stuart and Anthony Aveni (who is also an astronomer), have quite clearly stated that we have no proof that the Maya knew about precession.[2] This is because it is not explicitly referred to in any of their books or monuments.

But, to be honest, even they would have to admit that only 5 books of the Maya people have survived, out of potentially thousands of books that they wrote. So, who knows?

There are a lot of things not explicitly mentioned in the few existing Mayan books, like what their principle food crop was. Based on the surviving Mayan texts, you might think it was maize, but many archaeologists now believe that manioc was a more common staple of their diet. So it could be that they knew about precession, but it wasn't that important in their day to day affairs. After all, if it only moves 1° in 72 years then it really isn't a practical concern when predicting eclipses or deciding when to plant crops.

There are some reasons to believe they did know about precession. While it is true that 1° per 72 years is too slow to make any difference in the vast majority of astronomical observations and computations, it might be just right for tracking something that only occurs once in thousands of years. And did the Maya track any such events? Yes, as it is written in the Popol Vuh, their mythology, they observed periodic global cataclysms of fire and flood, on a very long cycle, at least many thousands of years. But, is there any other reason to believe that they connected these cycles to the precession of the equinox? Yes.

The Long Count, the longest period of time in the Mayan

calendar, is 5200 Mayan years. And the Maya made a point of celebrating anniversaries that were divisible by 5. Well, what is 5 times 5200 years? The answer is 26,000 years, the same as one complete cycle of precession of the equinoxes. That is an odd coincidence. Still, it might just be a coincidence. Actually, one precessional cycle is really less than 26,000. Established science clocked it at about 25,920 years, and there is now evidence that it is even shorter than that. Interestingly, you will note that the Mayan year is only 360 days. So 5200 Mayan years really are less than 5200 seasonal years, it's about 5127 years. That means that five Long Counts of Mayan time is actually very close to the exact amount of time that it takes for the Earth to complete one precessional cycle. So there is some evidence, however circumstantial, that they knew of precession.

Galactic Alignment

You will remember that some of the Mayan prophecies, mentioned in the last section, refer to the "mouth of the well". And, there are other odd references. Well, this might refer to the dark rift in the center of the Milky Way. Why is that important? The last Mayan Long Count ended exactly on the winter solstice, December 21, 2012. And, if you track the movement of the Sun against the stars on every December 21st, then you will note that the Sun has been moving into that dark rift for hundreds of years. On December 21st of 2012, it lined up with the very center of the dark rift, which we now know is the very center of our galaxy. This is known as the galactic alignment.[3]

Now you might be thinking that since nothing happened in 2012, then why should we care? The prophecy is false, right? The answer is that it takes over a century for the Sun to transit accross the center of the galaxy. The prophecy may hold true for the basic observation, but not the exact date. In other words, so long as the Sun appears to be in the "mouth of the well" on the winter solstice, we are potentially in danger.

> "At the mouth of the well, Uuc-yab-nal, it is
> established… It shall dawn in the South"
> -Chumayel Prophecy #2

The term "mouth of the well" could refer to the dark center of the Milky Way. Or, it could literally be a water well, in the middle of the town of Uuc-yab-nal, what we call Chichen Itza. Most scientists will tell you that this means nothing, even if it is referring to the Sun appearing in front of the center of galaxy. It is merely our perception from the Earth of the Sun relative to an arbitrary fixed point in the sky. There is no force-field unleashed by the galactic center, just because our Sun appears to be lined up with the galaxy. Others disagree, convinced that the Sun and planets lining up with the center of the galaxy could indeed trigger a global cataclysm. But I think people on both sides, the skeptics and the kool-aid-drinking apocalyptic devotees, completely miss the point on this issue. Let's look at the relationship between precession and prophecy a little closer.

Precession is in The Eye of the Beholder.

There is an important feature that I should explain here, for those who don't fully understand precession. It might surprise many people that our perceived "location" in the precessional cycle is really quite arbitrary. To put it another way, every year the Sun appears to be against a backdrop of different constellations in different months. This is why it is sometimes called "solar precession". We all know this from astrology. As mentioned above, if you are born in, say, late February or early March, you are a Pisces. That means that at the time of year when you were born, the Sun is right in front of the constellation Pisces (actually its not, because of precession they are now off by a month). What's important here is that the sun crosses in front all the constellations, and that includes the center of the galaxy, every year! So, the galactic alignment actually happens every year! And it always has.

The only thing that changes from one millennium to another is the time of year in which the galactic alignment occurs. Because it changes very slowly, the galactic alignment has been occurring on the winter solstice every year, for a number of years now. The winter solstice of 2012 was special only because the Sun appeared to be in the very center of the Milky Way on the exact date that the Mayan Long Count ends. But as I've said, this alignment happens every year, the only thing that changes is the time of the year that it occurs.

Interestingly, the precession of the equinoxes is usually observed on the spring equinox, and not usually on the solstice at all. That's why it is more often called "precession of the equinoxes". According to Walter Cruttenden, in ancient times they did not even measure precession on the spring equinox, but rather on the autumnal equinox. It was not until the depth of the dark ages that this information was lost, and people began mistakenly measuring the precession on the spring equinox.[4] So, according to the ancient zodiac, we're not entering the Age of Aquarius at all, we are really entering the Age of Leo.

By all this discussion of what you might call "precessional relativity", I think you can see that the individual constellations of the zodiac are indeed quite arbitrary. And, for that matter, so is the galactic alignment. There are interesting scientific and mythological aspects to the phenomenon of precession, but what constellation or galaxy that the Sun appears to be in front of, from our perspective on Earth, is not one of them. It is, scientifically, the least interesting thing about precession. The skeptics are right on this issue. There is nothing magical about the Sun appearing to be in front of the center of the galaxy. There is only one potentially important aspect of monitoring what stars the Sun is in front of, and that is if we use it as a time marker. Let me explain.

Suppose that I knew, from National Weather Service bulletins that a class 5 hurricane was going to hit at about 3 pm. Well, there's nothing magical about the time 3 pm. In fact 3 pm occurs every day and usually there is no hurricane. There is no special importance to the number 3, or to the Sun being located in the South-West sky either. In fact, if it is daylight savings time, it might not really be 3 pm at all! It could really be 2 pm. However if I wanted to warn someone, call them or leave them a note, then I'd definitely want to tell them that the storm is due to hit at 3 pm. If they have a clock or a watch, they can at least prepare themselves, board the windows, get in the basement or whatever, and they know exactly how much time they have, and when to brace themselves for the storm.

Now think of solar precession as just a great big clock in the night sky. If the ancients wanted to warn us about a reoccurring natural disaster that is due to come back in thousands of years, then a clever way to do it would be to leave a message that says something like, "be careful when the Sun is in the center of the dark rift, on the winter solstice." This is precisely the message that John Major Jenkins and others believe that the Mayans were trying to send us. Not "look out for the center of the galaxy because when the Sun goes near it a cosmic beam is activated." That doesn't really make sense scientifically. But, using the precession as a time marker makes a lot of sense. That is actually a smart and easy way to send a *message in a bottle* from across thousands of years of time.

As I said, the Sun will be lined up with the center of the galaxy (dark rift) for at least another 50 years, so even if we passed the midpoint, the prophecy may still hold true. What's important here is the basic idea of an ancient prophecy using a precessional time marker. That concept is scientifically quite sound. If this was all there was to precession, it would be enough to consider important, but there may be even more to it than just that.

Changes in Precessional Theory

There are some odd things about precession that scientists have discovered recently. There are anomalies that just don't make sense. In Walter Cruttenden's book, *Lost Star of Myth and Time*, he brings up some interesting facts. He makes the following observations:

> "As we know, Newton's equations never did match observed precession rates, so along came d'Alembert (1717-1783), followed by many others who have continually tweaked the formula to match observation. Ironically none questioned the underlying theory (in science, you usually don't question Newton). And so no one has stood back to ask if this 'wobble' might just be an apparent

motion, one not occurring within our local reference frame of the solar system."[5]

He goes on point out that "there is no evidence that this change (precession) in the spin axis occurs relative to the Sun, or Moon, or Venus, or anything else 'within' the solar system; this means that it may not be caused by 'local' lunisolar forces." What he is pointing out is that, up until recently, it was believed that precession was caused by the gravitational pull of the Moon and the Sun on the Earth, which causes it to 'wobble', just as a top does. That is called the *lunisolar* theory of precession. But, careful observation has revealed that the Earth is not wobbling relative to the Moon or Sun, or Venus, or other planets, only relative to the stars outside our solar system. Cruttenden continues:

> "The big thing wrong with this whole dynamist approach (the process of looking strictly at the local dynamics) is the assumption that the Earth's axis wobbles relative to all objects inside or outside the solar system. ...But, the truth is the so-called 'wobble' is primarily the geometric effect of an unknown motion. There is an unaccounted-for reference frame – the solar system curving through space – producing the phenomenon we call precession."[6]

Cruttenden goes on to cite researchers Karl-Heinz Homann and Uwe Homann, who have looked at this situation. They conclude that "the 'theory' of lunisolar precession mechanics simply do not fit the actual observed motions of the Earth moving through space."[7] He cites mathematician Eugen Negut, who has stated that precession cannot possibly display the dynamics of a spinning top, the common analogy that we used above. He argues that there must be another force acting on not just the Earth, but the entire solar system. Cruttenden reports the fact that "precession is actually accelerating and acts more like a body that follows Kepler's laws (in an elliptical orbit) than a wobbling top that should be slowing down."[8]

That's right, scientists have discovered that the precessional cycle is speeding up. If you measured it 1000 years ago, it might have appeared that it took just under 26,000 years to complete one cycle. But over the last century it has begun to speed up. In fact it is now predicted to take just 24,000 years.[9]

This new finding is odd on many different levels. First, it means that if the Maya knew of precession and linked it to the Long Count, then they were wrong. So, if they were using the Long Count (LC) to measure precession - such that 5 LC = 1 precession cycle - then they were off by 2000 years. They measured it just as we did, and fairly accurately, except that they had no knowledge of its future acceleration.

Secondly, this new information validates the prophecy made by the Vedic scriptures thousands of years ago. You will remember they are among the oldest sacred texts in the world, and they believed humanity oscillates between 12,000 years of progression, and 12,000 years of regression, taking 24,000 years to complete a whole cycle.[10] Yet another striking coincidence?

Finally, and most importantly, this new finding is very unusual because it violates a basic law of nature. Objects in motion tend to stay in motion, unless acted upon by a force. That means, all things being equal, that the rate of precession should not change, unless acted upon by a force. If anything, it should slow down as it loses momentum, as Cruttenden points out above. What is the force that could be making precession speed up? Good question. Scientists are not sure. It does appear to be speeding up at a rate similar to that of an object moving in an elliptical orbit. That is a clue, but not an explanation.

You will see how this interesting finding about Precession is an important detail to remember later. Now we are ready to begin our investigation into the scientific theories of a looming cataclysm.

Einstein & Precession

By the way it should be noted that it was the issue of precession that led to the first evidence in support of Einstein's theory of Relativity. There are different types of precession and so far we have only referred to the axial precession of the Earth. Another type, called Perihelion Precession, measures the deviation of a planet's orbit around the Sun. Newtonian physics (standard physics used before Einstein's relativity) predicted the rate of precession for planets such as Mercury. But, Mercury deviated from the amount of precession that was predicted by Newtonian physics. This was a puzzle to astronomers and physicists at the time. Einstein's equations, however, were able to more accurately explain and predict the precession of Mercury, and this was the first evidence found supporting Einstein's theory of relativity, which led to it being accepted by the scientific community.[11]

7

EARTH CRUST DISPLACEMENT THEORY

Probably one of the most well known and popular apocalyptic theories of planetary destruction, involves the earth crust displacement theory of Charles Hapgood, popularized in the best selling book, *Fingerprints of the Gods*, by Graham Hancock, and used to dramatic effect in the movie *2012*.[1] I feel that I should point out that this is, indeed, a serious scientific theory, though it has some serious flaws and most scientists still have reservations. The thing that is so intriguing about this theory is how it came about.

Charles Hapgood taught the history of science at Keene College in New Hampshire. As part of his research he was studying an old map, the Piri Reis Map, dated to 1513. What astonished Hapgood was the inclusion of the continent of Antarctica, which would not be discovered until 1818. He spent the next several years studying the geography of Antarctica, the science of map-making, and history of who drew up this map and what source material they used. The results were even more astounding.

He found that the coastline of Antarctica that the map revealed was the actual ice-free shoreline. But this part of Antarctica has not been ice-free for 4000 years. This was a map of the coastline of Antarctica as it looked 4000 years ago, before it was completely covered by ice.[2]

He then studied other early maps such as the Oronteus Finaeus Map (1531), the Mercator Map (1569), and the Bauche

Map (1737). After studying the details of many ancient maps, he found other anachronisms such as the inclusion of the Falkland Islands, before they were supposedly discovered. Also he found maps with estuaries where great rivers would eventually flow, and maps with an ice sheet over northern Europe, as it looked in the last Ice Age, and so on. The Oronteus Finaeus Map showed the South Pole in a different location. And, some of these ancient maps used cartography techniques that would not be invented for hundreds of years. They employed spherical trigonometry, and they had an accurate measure of longitude, which would not be perfected until the clock was invented.

The history of the maps indicate that they had been drawn from source maps that may well date back thousands of years. Most amazing, the Bauche Map not only showed the continent of Antarctica 80 years before it was discovered it showed the center of the continent, the very South Pole, completely free of ice, with a channel that cuts through it. We now know that this *is* how the land mass of Antarctica looks beneath the mile-thick layer of ice. But how did they know that thousands of years ago?[3]

Without any theory of how these anomalies could possibly exist, his work was largely ignored. Unfortunately, in science, when you present evidence that does not fit the existing scientific models, the evidence is usually just ignored. The prevailing wisdom was that there could not have been any advanced civilizations, which mapped the Southern Hemisphere, 4000 years ago. The idea was so preposterous that few even bothered to look at the evidence. Of course, since then a glut of archaeological evidence has caused us to at least begin to rethink human history, prior to 4000 years ago, and loosen our grip on beliefs that we once took for granted.

Hapgood spent the next years intensively studying this

perplexing mystery of history and science. The more he learned about climatology, and climate data from the last ice age, the more the mystery deepened. He discovered that while North America was in the midst of an ice age, at the same time parts of Siberia were significantly warmer than they are now.

Finally, he happened onto one explanation that neatly explained a number of anomalies at once: crustal displacement. What if the lithosphere, that is the Earth's entire crust, could shift about on its core? After all, the crust is sort of floating on top of a molten core, much the way volcanic rock floats on rivers of lava, from volcanoes. This would mean that not only does the magnetic north pole move from time to time, we know that, but that the actual land mass shifts positions.

This is independent of plate-tectonics; were not talking about one seismic plate moving against another. We are referring to the entire crust of the Earth shifting. So what once was the temperate South Atlantic, is now the frozen South Pole. This movement might produce massive earthquakes around the world, but that is a separate geological action, secondary to crustal displacement.

This simple idea would explain why Antarctica was not always under ice, and why North America *was* once under a mile thick sheet of ice (the last ice age), at the same time. This solves one of the most debated and unresolved issues in science, which is the exact cause of the ice ages. There have been many theories, but no one had ever conclusively resolved this problem.

Hapgood found early support and collaboration from an unlikely source, Albert Einstein. Yes, that Einstein. In a letter to Hapgood, he wrote:

> "I have read already some years ago in a popular

article about the idea that eccentric masses of ice, accumulated near a pole, could produce from time to time considerable dislocations of the floating rigid crust of the earth. I have never occupied myself with this problem, but my impression is that a careful study of this hypothesis is really desirable."[4]

-Albert Einstein

He went on to write the forward for Hapgood's first book, where he wrote that he often gets contacted by people who wish to consult him concerning their unpublished ideas. Dr. Hapgood's idea, however, was different.

"The very first communication, however, that I received from Mr. Hapgood electrified me. His idea is original, of great simplicity, and – if it continues to prove itself - of great importance to everything that is related to the history of the earth's surface."[5]

-Albert Einstein

As you can see, even from the beginning, there were noted scientists who responded well to his theory. It cannot be underestimated what a problem the cause of the ice age was for scientists. Ironically, the concept of crustal displacement was thought of much earlier, but immediately dismissed.

"Scores of methods of accounting for ice ages have been proposed, and probably no other geological problem has been so seriously discussed, not only by glaciologists, but by meteorologists and

biologists; yet no theory is generally accepted. …
Unless the continents have shifted their positions
since that time (ice age), the permo-barboniferous
glaciation occurred chiefly in what is now the
southern temperate zone, and did not reach the
arctic regions at all."[6]

-A. P. Coleman, Ph.D.

Even 25 years after the above quote was written, nothing
had changed. Professor J. K. Charlesworth of Queens College
wrote that "the cause of all these changes (ice ages), one of the
greatest riddles in geological history, remains unsolved; despite
the endeavors of generations of astronomers, biologists, geologists,
meteorologists, and physicists, it still eludes us."[7] The biggest
problem for researchers was a simple riddle that they just could not
solve.

During the last ice age, why was it warmer in the polar
regions, at the same time that North America was experiencing
an ice age? It was a simple question, why was it warmer in some
northern latitudes, while it was cooler at some lower latitudes? A
major ice age would affect the whole planet, wouldn't it? And the
pattern was not consistent; in fact, it was rather idiosyncratic. If
you go back far enough, you find that whenever one part of the
world is experiencing an ice age, another normally colder part of
the world was experiencing a more temperate climate with a rich
crop vegetation.

For instance, during the last ice age, Siberia, now a
frozen wasteland, was enjoying a much more mild climate. That
is something that just shouldn't happen, and does not make sense

any way you look at it, except for one theory, crustal displacement. And, in the 50 years since he first presented his theory, the above contradictory evidence has only grown stronger and more numerous. Fossilized forests were found in Antarctica, fossilized palm leaves found at the North Pole, and frozen Mammoths from the last ice age were found with a variety of fresh grasses and vegetation in their stomachs; and this was in northern latitude areas that are now completely frozen. No theory of volcanic ash or dust in the atmosphere producing what we would call a "nuclear winter" would account for such disturbances. Neither would increased or decreased heat from the Sun. In short, no conceivable theory could account for this evidence, except earth crust displacement.

Some theorists were very enthusiastic about Hapgood's theory, others were not. One of the major problems with this theory was the causal factor. This is something that Einstein himself worked on. The prevailing theory was that it must be the heavy ice cap, disproportionately weighted at the pole, which when combined with the centrifugal forces of the spinning of the earth might eventually dislodge the planet's crust. It would then slide around on the molten core. For those who looked at the theory, this was their biggest problem, and it was essentially a problem of math and physics. Einstein pondered this problem.

> "Without a doubt the earth's crust is strong enough not to give way proportionately as the ice is deposited. The only doubtful assumption is that the earth's crust can be moved easily enough over the inner layers. ... If the earth's crust is really so easily displaced over its substratum as this theory

requires, then the rigid masses near the earth's
surface must be distributed in such a way that
they give rise to no other considerable centrifugal
momentum, which would tend to displace the
crust by centrifugal effect."[8]

-Albert Einstein

In the end, the trigger mechanism, occurring at regular intervals,
and producing the kind of evidence that are noted above, has not
been found. It turns out that the numbers just don't add up. The
massive bulge in the earth's crust at the equator keeps the crust
firmly in place through the rotational forces acting on it. There is
simply not enough ice at the caps, ever, to pull the earth's crust off
its current placement. In the last edition of Hapgood's book "The
Path of the Pole" he admits as much.

"I am conscious of the many problems that remain
to be solved, especially that of the mechanism
of displacement. These are intensified by the
assumptions I have been forced to make regarding
the speed and recent date of some of these
displacements. Many difficulties, some real and
some fictitious, will confront this theory as they do
any new and far-reaching assumption."[9]

-Charles Hapgood

He goes on to propose what further research should
be done to test and prove or disprove this theory. Many of his
suggestions for further research have, for whatever reason, not been

completed. He wrote "Studies should be directed to the question of whether opposing centrifugal (or centripetal) effects of isostatic anomalies may not be factors in seismicity."[10]

Alternative Causes for Polar Shifts

Several scientists have tried to solve this problem and come up with an alternative causal mechanism for crustal displacement. One of the most compelling ideas has come from former Admiral, and retired NATO research engineer, Flavio Barbiero. He has checked, and vetted Hapgood's theory, and the supporting evidence for the theory, and he came up with at least one idea that could possibly be a trigger. He suggests that an asteroid, one kilometer wide, colliding with the earth at a speed of 20 kilometers per second (a very reasonable estimate) would produce an enormous blast of energy. It would, he writes, "liberate an energy equivalent to ten billion Hiroshima-type nuclear bombs."[11]

Further, if this occurred in the ocean, pushing the entire ocean in one direction, it could produce sufficient forces against the land masses to dislodge the crust. The equatorial bulge would simply adjust itself to regain homeostasis. Thinner parts of the crust would be sent out to the equatorial regions. And the previous equatorial bulge would likewise break apart through massive seismic activity, reconstituting itself in the new equator.[12]

The biggest problem with Barbiero's theory is that such an event would occur at random intervals, and the displacement size and direction would be equally random. But it's not. According to Hapgood's research, the movement of the poles has been very consistent from one event to another. In my own analysis of Hapgood's data I found that the earth crust shift always occurs with the same angle of direction, and about the same magnitude of effect, 25˙-30˙ latitude. It's unclear if Hapgood discovered this himself, but it does add credence to his data.

How is it possible that we keep managing to hit a large asteroid every 24,000 years, and it always knocks up in exactly the same direction, and with the same force? What is the mechanism there? And, if it is not a regularly occurring event, then how is it related to precession? Unfortunately, this theory does not seem any more promising than Einstein's theory of centrifugal forces acting on the polar ice caps.

I will present a summary of the evidence for and against the crustal displacement theory in a later section. For now suffice it to say that, except for the problem of the missing causal mechanism for crustal displacement, it is probably the most elegant theory so far that neatly explains everything we see in the ancient accounts of reoccurring global cataclysms. This one theory explains how periodic ice ages came about, it explains periodic natural disasters by fire (lava shooting up from the core, due to the enormous frictional forces), destruction by flood (enormous tidal waves produced by massive earthquakes, and the general sloshing of our entire oceans up over the continents). Hapgood's maps even document the existence of unknown and fairly advanced civilizations in prehistory, not unlike Atlantis.[13] All that is missing is the timing of it, which is directly related to the missing causal mechanism.

Had Hapgood only bothered to study Mayan prophecies he might have explored a whole new line of research involving the possible timing of these periodic shifts in the earth's crust, and hence he may have found the missing mechanism causing or facilitating this event. Perhaps there is a clump of asteroids that come close to the earth every 24,000 years. That would neatly tie in with the Barbiero theory. Perhaps they are undetected because they are on a very long, wide, elliptical orbit.

All we can say for sure is that if there was a way for the earth's crust to be displaced on a regular interval, especially if it is close to the cycle of the procession of the equinoxes, then that would definitely answer a lot of our questions. And, any skeptic would have to acknowledge that fact. Earth crust displacement, if real, would be the most elegant solution to a very long list of unsolved mysteries in science, up to and including the step-wise progression of evolution as an adaptation to changing environmental factors.

This theory explains how a cataclysm could occur, which would create violent geo-thermal events, volcanic eruptions, earthquakes, and devastating tidal waves. More important, this theory is the only one capable of scientifically explaining the unusually common mythological references to the Sun rising in a different direction, at the beginning of each new age.

This idea of a dramatic shift in the direction of the rising Sun is a bizarre and unique mythological theme. It seems hard to believe it was mere coincidence that so many cultures around the world would report such a similar and specific phenomenon.

Of all the credible scientific theories that I've looked at, this is the only one capable of producing the particular effect of apparently changing the path of the Sun. All that is missing is a trigger, a mechanism that would produce such a phenomenon.

8

JOVIAN MASS THEORY

The next most relevant scientific theory of a global catastrophe is the proposal of a Jovian Mass. This name is probably unfamiliar to even devotees of apocalyptic literature. I use the name "jovian mass", because it is the most scientifically accurate and accepted term, and carries with it the least amount of assumptions. The presence of an unknown jovian mass outside our solar system is a real scientific topic, which is presented at scientific conferences, published in respected journals, and debated among the top astrophysicists of the world. Of course, you may be more aware of the common vernacular for this concept, planet X, Nibiru, Mithra, the Lost Star, or simply the "Destroyer." So what is a jovian mass? The root word Jove is Latin for the Roman god Jupiter, or the planet Jupiter. So a jovian mass is simply a very large heavenly body, as big as or bigger than Jupiter.

It is important to note that this idea has been batted around for thousands of years, as shown in the mythology section. It is, however, a fairly serious, contemporary scientific question as well. Are there objects as big as or bigger than Jupiter orbiting our sun that we don't know about? All scientists would agree that there is nothing inside our solar system as big as Jupiter that we do not know about. But could one exist outside our solar system? That is a different story. The area outside of our solar system is believed to be a collection of orbital debris, referred to as the Oort cloud. We know that a great many comets exist on long elliptical orbits coming in and out of our solar system. Most of them, at any given time, exist outside our solar system, yet are still held in orbit by our Sun. What else might exist in the Oort cloud? Well dust, gases, ice, small rocks and frozen asteroids for sure. But, could something larger exist there? The answer is yes; it is theoretically possible. But what all is actually out there, we may never know.

Most astrophysicists think that it would be unlikely for there to be planet like Jupiter in the Oort cloud. Could it be a star? Maybe. That would mean that our Sun is actually part of a binary system. It is possible that either a brown dwarf or a black dwarf could be hiding very far out from our solar system. But, there are problems with each of these theories. A brown dwarf would be more likely to be detected. While, a black dwarf, a dead star, would have to be very old. It is estimated that it would take 100 billion years for a star to completely cool down enough to be considered a black dwarf. But it is also estimated that the entire Universe is only 13.7 billion years old. So black dwarfs could not possibly exist. Of course we don't really know, we can only speculate on these issues. Perhaps a smaller star could cool more rapidly. Maybe it doesn't have to cool down to the point of a black dwarf. It may avoid detection by simply being relatively cool, with little or no observable radiation, from this distance.

Theoretically, it could be possible for a very small star (still big compared to the size of our planets) to be lurking in the shadows, very far out in space. But, it would have to be far enough out that it could not be detected. That would mean that it would have a very, very long orbit, even tens of thousands of years long. Ironically, that perfectly fits not just the mythology of the ancients, but also the length of the procession of the equinoxes, which is 24,000 years.

History of Research on Binary Systems

Unfortunately, while it may be theoretically possible that we have a large jovian mass orbiting our solar system, that is not enough to constitute a sound theoretical model. After all, anything is theoretically possible, however improbable it may be. But you would need some compelling thread of logic to show why it is not only possible but also probable.

Fortunately, astronomers' attitudes about binary star systems in the galaxy has progressed in a very definite direction over the last one hundred years. Binary systems have gone from something believed to be very rare to being considered quite common. George S. Boyd, a contemporary of Einstein, was one of the first astronomers to propose that our own solar system was in fact a binary star system. In his monograph, published in 1937, he argued that the original mass of our solar system must have "divided into two separate stars, or binaries."[1] He went on to say that "double stars are actually known by tens of thousands. Probably one star out of every three or four in our galaxy is a double or binary."[2]

Ironically, the estimated number of binary star systems has only gone up since Boyd published his work. It's now believed that two out of three stars are binary, and it may be even more common than that. Boyd, concluded by saying that since we obviously do not currently have a binary star system, it must have either rejoined with its twin, or perhaps broken up to become planets. Jupiter, may be what's left of our binary, a star that wasn't big enough to ignite. This is an idea that has persisted among astronomers and astrophysicists ever since.

Einstein was obviously aware of binary systems, and studied their existence in relation to his theory of relativity. In fact, he made certain predictions about binary systems which have only recently been proven true. For instance, a recently discovered binary system with a large neutron star (twice the mass of the Sun), orbited by a small white dwarf has revealed itself to be the perfect laboratory for proving Einstein's theory.[3] There are a few alternative theories about the decay rate of the white dwarf, but it turns out that Einstein's theory best explains the data. The radio observations were so precise that they have been able to measure a change in the orbital period of 8 millionths of a second per year, exactly what Einstein's theory predicted. Unfortuneatly, he did not predict that our Sun had a binary twin. That was for more recent astronomers.

Walter Cruttenden, an archaeo-astronomer and Executive Director of the Binary Research Institute, has researched this binary theory as much as anyone and he has made a compelling case for the idea that our Sun should, in fact, have a binary twin. He makes his theoretical case based on the relative mass and angular momentum of objects in the solar system and the law of conservation of momentum. He presents in his book *Lost Star of Myth and Time*, the following theoretical argument:

> "The solar system was thought to have formed about 4.6 billion years ago out of a swirling cloud of gas and dust. As things slowed down and the planets congealed, angular momentum should have remained proportional to the mass of the objects within the solar system, according to the most accepted laws of physics. It

is a fact that the Sun contains the most mass of the solar system (estimated at 99.99%), which means under current solar system formation theory, it should also have most of the system's angular momentum. The problem is that it has less than 1% of the total angular momentum. Oops!"[4]

Cruttenden goes on to frame this problem in astrophysics as conundrum commensurate with that of geologists trying to explain the ice age. He goes on to explain the so-called solution that the leading scientists were forced to come up with.

"For years this was perhaps one of the best known and most discussed solar system anomalies because it was long thought that objects could not lose their angular momentum (by the law of conservation of angular momentum). It frustrated solar system theorists to no end, so recently scientists hypothesized that the Sun's angular momentum must have 'disappeared.' … Basically, it was there as required by our current understanding of physics, and then it (just) disappeared."[5]

He goes on to document many far-fetched theories of how and why the Sun's angular momentum might have disappeared, but none are very cogent or elegant. Finally, he argues against this "disappeared" theory by citing a Scientific American article, The Secret Lives of Stars, which explains how some young stars might eject matter (a key part of explaining loss of angular momentum) and that "most of the matter would end up being accreted, [and] some 10% might be ejected." So, Cruttenden concludes, "this 'disappeared' explanation does not really fit our particular Sun, that has somehow lost 99% of its angular momentum (while the planets have lost none of theirs)."[6] He goes on to make a key observation. "This does not fit observation, if that observation does not recognize the solar system as a moving reference frame." This is a key point because he argues that it is only when taking into account the motion of the solar system about another binary system that this anomaly is completely resolved. Cruttenden writes:

> "The binary theory offers a simple solution to the problem. The Sun's angular momentum is still *there*. It never went anywhere, and it is still proportional to its mass. Just as we calculate the planets' angular momentum based on their spin and orbital motion, so should we calculate the Sun's based on *its* spin and orbital motion. But if the Sun's only locally recognized orbital motion is in a small circle (caused by the gravitational pull of Jupiter's orbit), or if it is strictly around the center of the galaxy, then we have the angular momentum problem. But …see what happens if we include the Sun's motion in a 24,000 year binary orbit. Viola! It was there all the time!"[7]

He points out that in a binary model most of the Sun's angular momentum is in its movement through space in a binary orbit, just like the planets. A very large (24,000 year long) binary orbit takes up a lot of angular momentum, in fact, *the exact amount* predicted by the relative mass of the Sun. Now, that is one hell of a coincidence! But, there's more.

As Seen From Space

Remember in our discussion of the precession of the equinoxes, we pointed out an odd anomaly? The wobble of the earth's axis, which produces the effect of precession, doesn't seem to apply to any objects within the solar system, only objects outside of the solar system. So, it's as if the entire solar system is wobbling, not just the Earth. And, this wobble takes 24,000 years to complete a cycle. Well it turns out that one of the ways in which astronomers detect binary systems is to look for a wobble in the movement of a star system. These are called astrometric binaries. To an astronomer, this wobble indicates that the star may be moving around another invisible yet massive object such as a binary star, or what we might call a jovian mass if we didn't know what it was. This is especially true when one of the stars is too faint or small to see or emits no electromagnetic radiation, like a neutron star.[8] The wobble is one of the only clues to a binary system. Ok, now lets put it all together.

If it is not just our Earth but our entire solar system that is wobbling on a 24,000 year cycle, and if some alien race had been observing Earth over the last 25,000 years, what would they conclude? The answer is obvious to any astronomer, they would have to conclude that our Sun is part of a binary system, on a very, very long orbit around its companion star, an orbit that is 24,000 years long to be precise. So, from a distant star, our solar system looks exactly like a binary system.

This is such a simple and elegant theory that short of actually finding this binary companion, there is not much more to say. And, of course, if this object is only a few times the size of Jupiter, and if it is very far out in the Oort cloud, then finding it might prove quite difficult. If it is giving off little to no radiation, and too far out to reflect our Sun's radiation or light, and especially if we do not know where in the sky to look, then it could go undetected for a very long time.

The Dog/Pet of Our Sun God

Cruttenden goes out on a limb, based on his reading of ancient mythology, and suggests that our binary star is actually visible in the night sky. He has proposed that our binary star is none other than Sirius, the dog star, the brightest star in the night sky. "The more one looks at the evidence the more it appears that Sirius, one of our closest stars, and the absolute brightest star in the sky, could be our partner Sun."[9] Cruttenden was actually not the first to propose this. The famous Egyptologist and philosopher R. A. Schwaller de Lubicz (1887-1961) first proposed that Sirius might by a companion star to our Sun, based on his reading of ancient Egyptian calendars and mythology. He found that when you plot the movement of Sirius in contrast to precession, over thousands of years, you see that our Sun and Sirius are actually moving toward each other.[10]

Though it seems unlikely at first blush, it is true that Sirius is on the move, and over thousands of years it is predicted to cross the night sky. And, remember that in ancient Sumerian Nibiru means "crossing", perhaps meaning the "crossing star". Still, one could argue that our binary companion would more likely be found somewhere in the Oort cloud surrounding our solar system, not nearly 9 light years away.

One thing in favor of this idea is that 24,000 years is a long time. If Sirius is our companion, then I clock it as moving at an average of 56,000 miles per hour. That sounds like a lot, but actually the Earth is moving around the Sun faster than that every second of every day, at about 66,000 miles per hour. So with an elliptical orbit of 24,000 years, at about 8.6 light years away, it is possible that Sirius actually could be in orbit with our Sun. Whether that makes sense given the gravitational forces involved is another matter. Cruttenden thinks so. And he has gained at least some support in the scientific community.

The good news about this proposal is that we know that we are nowhere near Sirius at the moment. According to Cruttenden, the Indian Vedic scriptures probably got it right. Remember they foretold of an ascending 12,000 year cycle, followed by a 12,000 year descending cycle, completing a 24,000 year long orbital cycle, which they believe actually causes the phenomenon of precession. According to Swami Sri Yukteswar, we passed the apogee of our orbit over 1500 years ago, and now we are headed back toward our companion star.[11] Cruttenden's calculations regarding the current acceleration of precession, confirms that based on Kepler's law we probably did pass the apogee point about 1500 years ago. That means is that if we began our ascent in around 500 AD as Swami Sri Yukteswar said, then we have another 10, 500 years before the next rendezvous with our companion star and a possible apocalyptic scenario. And, that is an encouraging thought.

9

OLIVER REISER &
A SYNTHESIS OF THEORIES

You will notice that we are only reviewing three scientific theories, even though there are many fascinating theories, which might explain the possible causes of apocalyptic doom. Remember that we're only interested in theories that would explain the specific mythological accounts and prophecies that were presented in the first section. That narrows it down quite a bit.

Sure, any cataclysm could produce destruction by fire or flood. But not many could account for both fire and flood, as well as something like what we might call a "nuclear winter" blotting out the Sun and causing a burning rain of black resin, not to mention the Sun rising in the wrong direction when it's all over. And of course all this will be caused by or coincide with a large heavenly body coming close to earth, on a regular orbit of 24,000 years. That's a fairly specific scenario.

These scientific theories of earth crust displacement combined with a binary jovian mass orbiting our Sun are the closest things I could find to explain the bizarre mythology our ancestors left for us. These scientific concepts, each conceivably plausible in their own right, are a distillation of all the hundreds of ideas, concepts, and scenarios that I've reviewed for this book. But I think these two might just be enough, with one possible exception, the theories of Oliver L. Reiser.

Oliver L. Reiser

There is one more theory, or perhaps I should say theorist, that I think might be important to introduce. Oliver Reiser (1885-1974) was a Professor at the University of Pittsburg, and similar to Charles Hapgood, he taught the philosophy of science. He published many books and scores of articles in his lifetime, and also like Hapgood, he corresponded with Albert Einstein on his ideas, and received help and collaboration from Einstein.[1]

It is hard to characterize the ideas of Dr. Reiser. His work defies description. He was a prolific writer, and decades if not centuries ahead of his time. Many of his theories have proven true. He accurately predicted the location of the Van Allen radiation belts. He believed that our magnetic poles had reversed periodically throughout history, and that we would find evidence of magnetically reversed layers in rock samples, this later proved to be true. He believed that the brain was a network of electrical signals, which is sensitive to ambient electromagnetic fields. Then he also connected these two phenomena and said that fluctuations in the earth's electromagnetic field, especially magnetic pole changes or reversals, could stimulate or alter human consciousness; that is still being researched, but the evidence so far is very interesting.[2]

He believed that these periodic magnetic pole reversals were related to cyclical changes in the Sun and the galaxy, and somehow connected with the precession of the equinoxes. He believed that during these periodic events biological mutation of the species would occur due to increased levels of cosmic radiation, acting on DNA through magnetic resonance. He connected this with Darwinian evolutionary theory, and the idea of punctuated equilibrium, which explains the sudden appearance of a new species.[3] This also produces the kind of step-wise progression of human evolution seen in the anthropological record.

Conceiving the Internet

Some of his ideas were very far out, yet ironically many of these too have proven true. He foresaw a future where there would be a global network of information and communication technology. He envisioned a world in which, not only could anyone anywhere on the earth talk to anyone else, but all people everywhere could access the entire wealth of human knowledge, by plugging into a collective bank of knowledge. He wrote about that idea 50 years before the first prototype of the internet was even invented. At the time, he had theorized that it would be some kind of global radio-television hookup. Even that was ahead of his time, since television was still a fairly new invention. He described this global network as:

> "a super-dispersive ether to the extent that ordinary velocity of transmission-of-information relations and inverse-square law of spatial relations are transcended in a higher manifold."[4]

He envisioned a world where distances did not matter. Anyone anywhere could communicate or interact with anyone else in real time, virtually instantaneously. Of course, now we take this for granted, but imagine these ideas were created by someone born in 1885, writing in the 30's and 40's!

The Electromagnetic Society

Other ideas were a little more dubious. He believed that the electrical network of our brains manifested as an electromagnetic field in or surrounding our head or body. He believed that the next step in evolution was for each person to become like a cell in a larger global organism. This could eventually create a planetary-wide electromagnetic field network of consciousness, he called the "Psi-layer". This "electromagnetic society" would be capable of connecting all of humanity in one vast telepathic network.

> "...the hypothesis that humanity is the emerging social cortex of an embryonic world organism suggests the possibility that there is a psychic ether (a sub-ether) which provides the medium for the functioning of constituent waves as archetypal forms. If there is a 'psychic ether' for a universal morphogenetic consciousness, could it be that there is a Psi-layer surrounding the earth like the ion blanket, reflecting the images of mankind back to the earth creatures?"[5]

He was convinced that, one way or another, the future evolution of humanity would include parapsychology, psychic abilities, and global consciousness, what he called the psychic revolution. This would lead to the psychic unity of all humanity. The global, communications, and information technology that he talked about would serve as a necessary foundation to facilitate the next step in human evolution.

"The mutant personalities of the new humanity may then carry forward the progressing psychic revolution which will ultimately yield an electromagnetic society with gestalt patterns of symbolism transferable across the social whole. A global radio and television hook-up employing planetary semantography is the earthly foundation for the coming psychic unity of mankind."[6]

If you just replace the words "radio and television" with "online" and replace "planetary semantography" with "a common language", you would get the following statement *A global internet hook-up employing a common language is the earthly foundation for the coming psychic unity of mankind.* He wrote that more than 50 years ago yet, as you can see, much of this has already started to happen. Not only has the internet created the kind of technological information and communication network he envisioned, but there have been changes in language as well. Out of thousands of languages that existed just 100 years ago, now there are only a few languages that dominate the internet, world business, world trade, airline communications, and so on. Of them all, English is quickly becoming the most common universal language.

Cosmic Humanism

He believed that the next step was galactic consciousness, spreading out to connect with other life-forms throughout the cosmos. He reasoned that we needed to create the kind of structural and societal changes necessary to facilitate this advance in human evolution as well. Not only does this include creating a global communications and information technology, but also creating a truly universal language capable of communicating with other intelligent life throughout the galaxy, possibly using mathematics corresponding to sounds, as was presented years later, in the movie *Close Encounters of the Third Kind*.

He also advocated a new type of global, spiritual, and ethical philosophy that is compatible with science. He argued in his book the *Integration of Human Knowledge*, that this new type of ethical humanism was needed to replace all the differing, primitive religious beliefs of the world, which keep us divided, are incompatible with science, and do not provide a sufficient common ground for humanity to share with each other, or with the rest of the cosmos.[7] Albert Einstein was intrigued and inspired by this idea. It was Einstein who named Reiser's new philosophy Cosmic Humanism.[8]

Radio Eugenics

Strangely enough, Reiser was ultimately more interested in eugenics than spontaneous evolution due to natural cataclysms. At the end of his last book, Cosmic Humanism, he goes into great detail about the proposed mechanisms and ethical objections to purposely creating human genetic mutations in order to create a super-human race of people. He proposed "a theory of radio-mutations based on the ability of rays of short wave length to produce changes in the genes and thus cause biological mutations."[9]

He goes on to say that natural evolution can take millions of years, and may not yield the results we would desire. "But how can evolution be speeded up?"[10] He concludes "we could, if we wished, subject human foetuses to irradiation by the by-products of atomic fission and fusion. ...Perhaps the beneficial use of atomic energy is a matter of selecting the right types and dosages (frequency and duration) of irradiation."[11] He didn't say whose babies he would experiment on, but he did indicate that we would have to get over our sentimental parental instincts.

This somewhat bizarre and creepy line of thinking deviates from our study of global cataclysms a bit, but it just goes to show how convinced Reiser was that delicate changes in the radiation bombarding human DNA could produce dramatic genetic mutations, and advance evolution. He simply didn't want to wait until the next pole shift, and leave it to chance. He wanted to use this scientific principle to design and genetically engineer the future of humanity.

The Last Puzzle Piece

How does this help to scientifically explain the ancient prophecies we studied earlier? Well, Reiser actually worked out many of the details of what we now know as the Mayan doomsday scenario, more 60 years ago. He predicted a connection between the precession of the equinoxes and some kind of periodic global event, involving a magnetic pole change or reversal, increased cosmic radiation, and a sudden evolutionary leap in the human species. The only thing he lacked was the Mayan prophecy itself. Mark Heley neatly summarizes the work and shortcomings of Reiser in the following:

> "Significantly, because of the time he was writing in,
> Reiser lacked the knowledge of two ideas that might have
> helped him complete his grand theory. These were

the winter-solstice galactic alignment identified by Jenkins
and knowledge of the Mayan calendar end date in 2012.
Reiser knew he still had missing elements, and in his
book *The Holyest Earth* he asks: 'What and where is the
master timing device (the sun-planet-galaxy clock) which
regulates the interdependent causal sequences to achieve
and maintain the astro-geo-bio-homo-social chain of a vast
interlocking and awsome teleology?' The Sun-planet-galaxy
clock may well be the galactic alignment and the master
timing device of the Mayan calendar."[12]

Heley goes on to write that "Reiser managed to be a 2012
theorist without ever even knowing about 2012".[13] I would go a
step further, and say that he was one of the most brilliant of all
doomsday theorists, because he did not reverse-engineer the event
from prophecies, but he predicted it in a virtual vacuum. In other
words, he did not start with the idea of a doomsday prophecy, or the
myth of Atlantis, and then try to figure out how it could happen. He
deduced the whole scenario based on his vast knowledge of science,
which led him to this unique theory as the only logical conclusion
to explain all the data from many different fields of study. That is
why his scenario is so conspicuously lacking the fires and floods and
other eyewitness-type accounts so common in the mythology.

If you think about it, his scenario doesn't really preclude
a natural disaster. In fact, it could well be very disastrous. But,
that just wasn't his focus. He was more interested in finding the
mechanism that might produce sudden bursts in the evolution
of the species, which we find in the fossil evidence and the
anthropological record. Perhaps this is what the ancients were
referring to when they wrote about the three or four previous

"creations" of humanity.

Reiser provides us something that the other theories are lacking, an evolutionary theory. The Mayans talked about the previous ages as a series of attempts to create an improved humanity. Many of the prophecies from around the world talk about how the upcoming new age of humanity will somehow involve a growth in spirituality or higher consciousness. The Vedic scriptures describe how we are ascending to become a more perfect being. The modern day Maya elders, say very much the same thing, that this new age will be a major step forward in humanity, affecting our DNA, causing us to evolve higher consciousness. Then there are the New Age prophecies of a coming brotherhood of Man, which will soon manifest in the Age of Aquarius. So far, however, no other scientific theory except for Reiser's has come close to explaining how this might all occur, much less connecting it in a meaningful way to either precession or pole shifts.

The Reiser evolutionary theory is like the missing piece of the puzzle. It provides the one thing that was lacking from the other scientific theories that we researched, while still validating core elements of the other theories. Interestingly, it does not compete with the other theories. In fact, all of the theories actually complement each other remarkably well. And, Reiser is the finishing touch.

A Synthesis of Scientific Theories

After I had sifted through all the different 2012 theories, ideas and concepts that I researched for this book, I have to admit that I was shocked when I saw what turned up. These theories neatly fit together like pieces of a theoretical jigsaw puzzle. It's uncanny the way each theory holds a missing piece of the other theories.

Not only does the theory of solar-system-wide precession suggest that we are part of a binary system; that would be impressive enough. But the idea that the speed of precession is accelerating is exactly what would be predicted by Kepler's laws, if our solar system were in an orbital path, heading back to a rendezvous with our binary companion. So, the revised processional theory actually validates the jovian mass / binary theory.

Then it hit me. Though no one had ever put the two together before, the jovian mass / binary theory actually validates the crustal displacement theory. Think about it. No one has ever come up with a credible trigger mechanism to create a crustal displacement of the earth's entire lithosphere, except for a large heavenly body crashing into the earth. Well, what if it didn't actually crash. What if a heavenly body, larger than Jupiter, passed close enough to us to produce changes in the earth? If a jovian mass moved past the earth, could it cause the entire lithosphere to slip about its core?

Of all the possible re-occurring trigger mechanisms, with a 24,000 cycle, a rendezvous with a binary companion would be the most likely. We can rule out an asteroid crash since there is no evidence that the earth has taken a direct hit by an asteroid every 24,000 years. But, a near pass of a jovian mass would leave precisely the kind destruction that we see reported in the mythology. This scenario, including both a jovian mass and crustal displacement, would involve a strong gravitational force acting on the earth, possibly capable of disrupting both the lithosphere and magnetic poles as well. It would occur at regular intervals of 24,000 years. It explains the mysterious "star" mentioned in the myths, and at the same time it also accounts for the unusual report of the Sun rising in a different direction. The two theories (crustal displacement and a jovian mass) combined with a revised precessional theory, creates a more solid and robust doomsday scenario than any one of them separately. Now add Reiser's theory and voila! You have an cataclysmic event linked to precession and pole shifts, affecting DNA to produce a new step in human evolution.

Now that we've reviewed all the theories and how they fit together, what kind of scenario does that suggest? Here's the scenario as we have it so far:

A Scientifically Credible Doomsday Scenario

1) Our Sun is in an orbital path around a jovian mass (an object several times larger than Jupiter), with an orbit of about 24,000 years. This orbital path produces an effect that we on earth perceive as precession. The rate of precession speeds up and slows down depending on where we are at in the elliptical orbit. The entire cycle takes about 24,000 years to complete.

2) Just before we rendezvous with this jovian mass, we start to see it as a new star in the sky, which gets brighter over time.

3) When our solar system comes swinging by its binary companion, the gravitational forces on the earth become more severe. At some point, the tangential gravitational pull on the earth's lithosphere is too much, and it separates from the molten core underneath, and begins to slip. This could happen rapidly, in days or weeks, or very gradually, over thousands of years, or it could be a series of small

but rapid shifts over a long period of time. The above theories do not specify this detail. Computer models that Hapgood did not have access to in the 70's could now help us answer these questions. But, one way or another, eventually crustal displacement occurs.

4) Such an earth crust displacement could easily cause massive geothermal energy to build up, resulting in volcanic eruptions, producing black burning ash to rain down, and a nuclear winter that would blot out the Sun for weeks or months. The breaking up and reconsolidation of the equatorial bulge would produce cataclysmic earthquakes of biblical proportions. Tidal waves up to a mile high would not be inconceivable. The climatologically implications are unknown, but it could conceivably rain for 40 days and 40 nights.

5) Following this cataclysm, the world is nearly destroyed; most of the human and animal populations are dead. Many species may become extinct. The entire crust of the earth has slipped from its former location. Perhaps there is a complete magnetic pole reversal. Cosmic rays are allowed to briefly strike the earth during a lapse in the magnetosphere, while the magnetic poles are fluctuating. Human DNA either resonantly or otherwise reacts to the cosmic rays, producing a new generation of human mutations. Hence, a new age of humanity is created.

6) The flood waters eventually recede. The volcanic activity diminishes, as do the aftershocks, and finally a new world emerges. The continents are reformed; perhaps the difference is only subtle, or it could be quite dramatic. As the former poles begin to melt, water levels rise again, and the coastal contours take their final shape. The new poles begin to freeze and accumulate

ice caps. Eventually, a new world emerges. The Sun dawns in a new direction, and there to witness it is a human child of the survivors, different from any human that has come before. A new version of humanity for a new age of Earth.

Conclusion

As Dr. Frankenstein said in the Mel Brook's film, Young Frankenstein, "It could WORK!" Even though it is fantastic beyond belief, this is actually a sound theoretical scenario. Of course, it's only a hypothetical scenario, and as yet we have not looked at any evidence to support these theories. But, at least theoretically, it hangs together as a plausible explanation for the type of global cataclysms we see described in ancient mythology.

I can't tell you the rapid progression of emotions I experienced when I saw all these different theories neatly coming together, each theory confirming and plugging holes in the others, forming a master synthesis of theories. I was excited at first, even triumphant. I had done it! I just solved one of the greatest mysteries of world. Suddenly, I realized what this meant. My heart sank. For the first time it hit me that this could really happen in my lifetime. Not so exciting.

To calm my nerves I reminded myself that we still have to look at the evidence. Then there is still the timing issue. If precession is accelerating, and if we are part of a binary system (a big if), then we are on the second half of our binary elliptical path. But, how far into the second half? We could have 12 months or 12 thousand years, we just don't know yet. We might not find any evidence to support these theories at all. We can only hope.

Part III

Scientific Evidence for a Coming Apocalypse

10

EVIDENCE OF PAST
CATACLYSMS

In the last two sections, we reviewed first the ancient mythology, and secondly the scientific theories related to the apocalypse. But the question remains, is there a shred of evidence that any of this is true? In this section we will look at all of the latest evidence that would support both the myths and theories.

We will first look at the mythological accounts of destroyed lost civilizations in the past. If the ancient accounts of past cataclysms are completely lacking any scientific support, that would make any prophecies for future cataclysms highly suspect. That is, if there is no evidence that the earth has ever suffered any such cataclysmic episode in the past, then why would we expect it to happen again? Also, if we can't corroborate their eyewitness accounts of the past, how accurate could they possibly be about events thousands of years in the future?

Some might think that this is not a necessary step. They may be quick to concede that civilizations have been destroyed in the past, and move on. But what many people may not realize is just how controversial these issues are. There are still many respected archaeologists who believe that civilization as we know it began no more than about 5,000 years ago. And that humanity never attained a level of sophistication or knowledge as great as even the Renaissance, in the previous four millennia. Most scholars believe that Atlantis was just a myth. So, proving this part of the mythology is not an empty gesture.

It would be scientifically important to establish the validity of the mythological back-story that previously unknown civilizations did indeed rise up and become great and sophisticated empires only to be destroyed by periodic cataclysms, and perhaps even sunk to the bottom of the ocean. If we can find evidence to support this, then the rest of the prophecy is not so outlandish after all. If they accurately recorded true accounts of long-forgotten history in their mythological stories, then perhaps we should listen to their dire warnings for the future.

We will start with one of the most famous and outlandish of all the antediluvian myths, Atlantis.

Revisiting Plato's Mythical Atlantis

Even now the official opinion of most experts is that Atlantis was just a myth, a fiction, nothing more. Though people have pointed out similarities between Atlantis and actual civilizations of prehistory, such as the Minoans, the consensus of expert opinion has always been that the myth of Atlantis was purely fictional. They do not even entertain the idea that the myth was partly based upon actual civilizations of the past. In the introduction of a recent translation of Timaeus and Critias, Andrew Gregory, a research fellow at Oxford, states the prevailing view that "there are some interesting parallels between Minoan Crete and Plato's Atlantis, but nothing compelling."[1]

We have already covered the myth of Atlantis as documented by the Greek philosopher Plato. To recap, he said that there was a great civilization, and that they were very advanced, with hot and cold running water, and very advanced technology.

He also said that the very large continent of Atlantis was through the pillars of Hercules, and beyond Europe. Actually the Rock of Gibraltar is all that's left of the pillars of Hercules, which were two giant rock pillars that stood on either side of the strait of Gibraltar in ancient times, connecting the Mediterranean sea and the Atlantic. So the mythical continent of Atlantis must be somewhere across the Atlantic, hence the name Atlantis.

There were other details of this civilization. Plato reported that the main city was on an island. A large canal of water encircled the center of the island. There was a cross-canal connecting the inner ring of water and the outer shores. He says that they had enjoyed an incredibly rich agricultural production. Plato reported that, "everything aromatic the earth produced today in the way of roots or shoots or shrubs or gums exuded by flowers or fruits was produced and supported by the island then."[2] Plato goes on to describe their acropolis, their buildings and statues, mostly of dolphins and flying horses, which the Atlanteans loved and included often in their art. He writes of the importance of worship of bulls in their ancient culture. He goes on to report their goddess worship and their mighty fleet of ships and other assorted aspects of a thriving international empire.

There are inconsistencies in Plato's account. He describes a circular island with a concentric ring of canals in one section, then describes a rectangular island rich in agriculture in another section. In another section, he also says that it was a continent larger than north Africa and Asia, and that lay across the Atlantic. So, which was it? Skeptics point to these discrepancies as proof that this was a made up empire, which never really did exist. It is not clear exactly what happened but Plato says that in a terrible day and night the civilization was wiped out by earthquakes and floods, and sunk to the bottom of the sea. But was that the circular island that sunk, or the rectangular island, or the whole continent? It's not clear. Presumably the whole continent, islands and all were sunk?

Atlantis Rising

In the last few decades more and more research has been done on the ancient Minoan civilization. Decades ago the apparent similarities between the Minoans and the mythical Atlantis were observed and discussed.[3] But recent evidence, however, has revealed just how advanced the Minoans were and how widespread their empire was, and it is causing even skeptics to take a second look at both the Minoans and the myth of Atlantis.[4]

For those not familiar with this civilization, the Minoan culture was first discovered on the Mediterranean island of Crete, in the early 20th century, by the British archaeologist Sir Arthur Evans. It was discovered that the civilization had flourished for several millennia on Crete, from about 5600 years ago to about 3200 years ago.[5] Crete is a large rectangular island in the middle of the Mediterranean, and as such it is one of the most fertile agricultural lands in all of the Mediterranean.

Among the many Minoan palaces and structures on the island, Evans discovered the Palace of Knossos. I'll never forget the first time I explored the ruins of Knossos. It would be impressive enough if it were from classical Greece, but it is much, much, older. Knossos is a very large structure, not unlike a multi-story apartment building, complete with dumbwaiters, hot and cold running water, indoor toilets, and dozens of beautiful color murals, many with dolphins or griffins. Evans had also discovered many depictions of

men dancing with bulls, and sculptures of bulls. Evans named the
civilization the Minoans, after the legendary King Minos of Crete.
The ruins were so well hidden because they had been completely
and utterly destroyed in some ancient cataclysm, and then buried
under a deep layer of mud and sand.

Starting in 1967, excavations at Akrotiri revealed the
best-known Minoan site outside of Crete.[6] Akrotiri is an ancient
Minoan city on the southern coast of the island of Santorini,
known in ancient times as the island of Thera, and lying about 68
miles north of Crete. The structures at Akrotiri were much more
completely destroyed than that of Knossos. I have to say that,
in spite of the level of destruction, while touring the excavations
at Akrotiri I could easily see features similar to that of Knossos.
It was a large complex structure with many individual units or
apartments. It was also a multi-story building like Knossos but the
upper levels had been obliterated, leaving only stairs to nowhere.
Fortunately a large mural was found nearly intact, with a depiction
of how Akrotiri looked more than 3500 years ago. It was indeed
very similar to that of Knossos, but even larger and more impressive
than anyone could have realized.

With regard to the myth of Atlantis, one of the most
interesting features about Akrotiri is the island itself. First, it is
a volcanic island, not unlike the big island of Hawaii. Secondly,
around 1450 BC it blew its top, not unlike Mount St. Helens, in
Washington. This left nothing but a ring of land around what was
once a very large volcanic island, and nothing but empty water in
the middle. To the visitors, coming to inspect the damage after the
terrible eruption, it must have looked like the bulk of the island
simply sunk to the bottom of the sea. No one could have witnessed

the blast and lived. And, no one at that time could imagine how an entire island could disappear, no matter how large the volcano. The current geography of the island still shows features of how it originally looked. There is currently a small island in the middle of the interior sea, with two natural canals connecting the outer sea with the inner sea. This center island within the island was there during the time of the Minoans, but it was much larger. This created the effect of concentric rings of land and water, as Plato described.

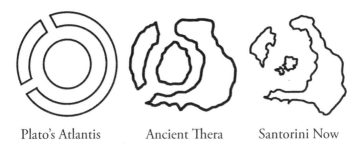

Plato's Atlantis   Ancient Thera   Santorini Now

Geologists feel certain that one of those canals was created by the volcanic blast, but the other, smaller canal was probably there before the blast. And, just as now, there may have been other canals. This smaller canal would have connected the outer Mediterranean sea with a ring of water that probably surrounded a center inner-island, which formed much the same way a center island is now in Santorini.

Also, it now appears clear that the island of Crete and all the Minoan cities and buildings there were destroyed at the same time by a tsunami over 100 feet high, created by the blast of the volcanic explosion, 68 miles away. It must have buried Knossos and much of the rest of the island in an avalanche of mud and sand, where it remained hidden for over 3300 years. And just as

Plato reported, this same tsunami would have destroyed the port cities of Greece in a terrible flood as well.

At this point, I am sure that you have noted enough similarities between Plato's Atlantis and the Minoan empire to see a connection. It is amazing to me that even with this information many skeptics argue that there could not possibly be a connection between the two. One issue that comes up over and over again is that Atlantis is supposed to contain a large continent beyond the pillars of Hercules, according to Plato. That means that it must lie somewhere across the Atlantic, not in the middle of the Mediterranean. They had an entire continent, not just a couple of islands. So where was this continent?

Discovering a Lost Continent

In a recent book published just two years ago, some astonishing new evidence has come to light about the Minoans. Gavin Menzies is a retired British submarine captain, author and researcher of naval history. In his book, *The Lost Empire of Atlantis (2011),* he cites some rather remarkable facts that he has uncovered.[7] Of course, he cites all the above obvious connections between Atlantis and the Minoans mentioned above, but he also reports some not so well known facts. As a naval officer, and scholar of ancient naval technology, he studied the archaeological finds of their ships, recovered Minoan shipwrecks off the coast of Turkey, and paintings of their fleets found on murals. He concluded that the Minoan ships were able to travel much further on open seas than previously thought. He also found evidence of extensive trade with Egypt, and also with the Balkan region, and even trade routes as far as

India, Spain and the British Isles. Perhaps because of his extensive naval background, he more than most was able to detail the sophistication and reach of the Minoan nautical capabilities. But he also reports some finds that just seemed like anomalies, prior to his research.

Menzies reports that the copper ore found in one Minoan shipwreck was 99% pure. But copper experts agree that there is only one area of the world that is known to produce ore of that grade, the Lake Superior copper of North America. He goes on to report the discovery of a tobacco beetle among the artifacts recovered from an archaeological dig at Akrotiri. But, this bug is indigenous to North America. Researchers had never been able to figure out how it got there 3,450 years ago.

He then presents research by a respected pathologist, Dr. Svetlana Balabanova, who studied bone, hair and soft tissue from nine Egyptian mummies, each about 3000 years old, and reported that they all showed evidence of cocaine and hashish usage, and most also showed use of nicotine.[8] What's so strange about this is that cocaine and nicotine are both exclusively found in American plants, cocoa and tobacco leaves. Subsequent research turned up evidence of nicotine in most of the Egyptian mummies studied.

Next, he found evidence of a man smoking a pipe in a painting on a Minoan sarcophagus, found at Hagia Triada.[9] While at the site, I had actually seen that sarcophagus and also thought it looked like a man smoking a pipe. But at the time I never made the connection. Ancient Greeks and other Mediterranean people did not smoke pipes, and should not have had access to tobacco. But obviously the Minoans did.

Probably the most astonishing evidence he uncovered was

the DNA evidence of the Minoans, showing who they were and where they came from. He cites research by Professor Constantinos Triantafyllidis of Thessaloniki's Aristotle University, who was part of a research group led by geneticists from Greece, the United States, Canada, Russia, and Turkey. Through DNA analysis, this research showed that the Minoans were closely related to the ancient indigenous people of Çatalhöyük in central Anatolia.[10] That makes sense, Anatolia was a thriving civilization more than 5000 years ago, with many of the same features. They too were a goddess-worshiping culture, with a reverence for bulls. They were located just inland of the Mediterranean, in what is now the southern part of Turkey. It would make sense that they would migrate to the nearby Mediterranean shores, and spread out to the islands over the centuries.

Interestingly, in all the world, there was just one other group that had the same DNA characteristics as the Minoans and the ancient Anatolians, and that is Native Americans! Menzies cites research by Michael Brown and Douglas Wallace at Emory University.[11] In their research the distribution of the Haplotype X2 indicated a connection with some indigenous tribes of North America. The epicenter of this connection seems to be in the Lake Superior region. In fact, as you spread out from Lake Superior, the DNA connection with indigenous people is progressively less, indicating that the Minoans might have inter-mingled gene pools with the Native Americans, specifically from the Lake Superior region. Why Lake Superior? Menzies concluded that they were probably there because Lake Superior had the richest, purest copper mines in the world. And it now appears that Minoan bronze was a chief export of the Minoan culture, along with tobacco, cocoa leaves and other assorted goods from America.

Finally, the most compelling evidence for archaeologists may be the large collection of Bronze Age artifacts found in the Lake Superior region. Archaeologists have long known about these artifacts and they attributed them to the "Old Copper Culture," presumably a Native American people who predated the indigenous tribes of that area.[12] The odd thing about the artifacts is their extreme antiquity as well as their technological sophistication. Many artifacts have been carbon dated to at least 3,500 years old, and some are much older. It also seemed unusual that the natives would have mined so much copper, up to a billion pounds, according to one estimate, then later stopped.[13] There is no dispute about the age of the artifacts, only their makers. The fact that these artifacts were inconsistent with other Native American cultures of that time period had been a bit of a mystery to archaeologists. But since the site was known to be inhabited for thousands of years by native peoples, who used flint tools and worked with float copper, it was assumed that they must have been the ones who mined the copper and made the artifacts. Now Menzies' research provides us an alternative scenario, even more consistent with the facts. He even goes so far as to do a side-by-side comparison of what look to be nearly identical artifacts from Minoans and from the Old Copper Culture of the same time period.[14] The artifacts provide compelling support for Minoans in the New World.

Menzies goes on to document other fascinating details about Minoan culture and its global reach, and other evidence of their voyages to the America. He even cited the discovery of cotton found in Old Copper Culture middens from the second millennium BC.[15] Of course mainstream historians dismiss his conclusions as fictitious, but they can't say that about the facts that he has uncovered. So far no one has been able to come up with one theory that explains all these strange anomalies, until Menzies. He has basically collected a wealth of facts that were mere curiosities, oddities, and anomalies, until now. But by showing that the Minoans were able to travel by sea much further than previously thought, he provides a compelling context for a whole collection of facts that were previously unexplainable. Now it looks like the Minoans might well have traveled to North America, over 1000 years before Plato wrote of Atlantis.

Why is this so important? Let's connect the dots. We now know that there were three principal areas of the Minoan empire. The first was Thera, now known as Santorini, a circular island, with a ring of water on the inside, connected by a canal with the outer island, which appeared to "sink" into the ocean, long, long before Plato's time. The second island was Crete, a large rectangular island that is very rich agriculturally, and contained many palaces and advanced cities, which were also destroyed (the same day). And now we know that there was a vast continent beyond the pillars of Hercules, across the Atlantic ocean, which was a colony and/or trading partner of the Minoans. It is reasonable to assume that while the sunken Minoan metropolis was Thera, and the palaces and agricultural center was Crete, the lost continent of Atlantis was America!

Validating Plato

The above research essentially validates that Plato got it right. It now appears that the reality of Atlantis was precisely as Plato described it. We just didn't know where to look, or how the locations were distributed. For many thousands of years, we didn't even know about the Minoans. Then once we knew of them, we didn't know that their capital appeared to suddenly sink into the ocean, or how that might even be possible. Once we discovered that, we still had no idea how extensive their nautical capabilities were. Now we see that they very well might have had an empire that stretched across the Atlantic, exactly where Plato said it was.

Of course, everyone conflated the two islands with the continent, assuming it was all part of one continent across the Atlantic. Even Plato probably made the same mistake. After a thousand years, even the Egyptian priest that Plato quotes may not have grasped the geography accurately. It is not until you put together all the pieces of the puzzle, that you see the simplicity of the whole picture. If it really was an exclusively American civilization, Plato never would have even heard of it, nor the Egyptians. Being the chief trading partners of the Minoans, the Egyptians would have known of the Minoans best, or the Keftiu as they called them, which fits what Plato said. Every piece fits together perfectly.

So, what does this mean for our investigation? Were the Minoans destroyed in 3,114 BC, the beginning of the last Long Count? No. Were they destroyed 24,000 years ago, exactly one precessional cycle ago? No. According to the Egyptian priest, quoted by Plato, Atlantis was destroyed about 9,000 years prior to relating the story to Solon, that would put it at about 11,500 years ago. But we now know that this date is not true either. If the Minoans were the fabled people of Atlantis, which it now appears likely that they were, then they were destroyed in a volcanic eruption in about 1,450 BC. So, the story is right, but the timing is way off. In addition to the timing issue, the biggest weakness of this find is that it was clearly unrelated to either crustal displacement, or a jovian mass. So it validates Plato's myth, but it doesn't really help support our scientific theories of global destruction at all.

What is really important about this research is that it simply shows that Atlantis is real. The lost continent of Atlantis has finally been found! I can't say how important this is to the antediluvian mythologists. Even if the timing is way off, it validates one of the most famous and storied myths in all of history, as being based on actual facts. Of course mainstream archaeologists will want to check and recheck all this research. Then they will probably begin to look for tobacco and cocoa leaves and other American items in tombs and artifacts from now on, something they never thought to do before. And it may be a generation before the textbooks are revised. Nonetheless, this adds additional confirmation to what was already a mountain of evidence linking the Minoans with the myth of Atlantis. So Plato was right all along. There was at least one very advanced civilization that we knew nothing about until the 20th century, with a continent across the Atlantic whose capital sunk into the sea.

Unfortunately, this still leaves us looking for evidence of other advanced cultures destroyed in a global flood, prior to 5000 years ago, when archaeologists say that civilization first began.

The Bay of Khambhat

One of the biggest finds in recent years to rock the archaeological world was the discovery of a previously unknown city, at the bottom of the Bay of Khambhat, (pronounced Cambay) in southern India.[16] An article by Indian geologist S. Badrinaryan published just a few years ago documents what was found while mapping the bottom of the bay.[17] Badrinaryan is listed as the chief geologist with the National Institute of Ocean Technology, an Indian government agency. He found not one but at least two previously undiscovered cities, one larger than Manhattan, lying at the bottom of the bay. His research uncovered artifacts of continuous human habitation going back 33,000 years. The twin cities discovered there appear to be inhabited as far back as 13,000 BC, and some "ill-fired" pottery found nearby was dated to more than 16,000 BC. These would be among the oldest fired pottery in the world. The well-fired pieces date to back to 7,000 BC. He feels certain that this culture is related to the Harappan civilization of the Indus Valley.

I remember studying the Harappans very well in my undergraduate archaeology courses. They were an amazing early civilization. According to archaeologist Robert Wenke, in his book *Patterns in Prehistory*, over 200 Harappan settlements have been discovered in the Indus Valley region of India and Pakistan.[18] He states that the largest is the lost city of Mohenjo-daro, covering more than 2.5 square kilometers, with a population of at least 40,000 people; it was discovered in 1922. It is among the oldest cities in the world. It appears to have been built around 4,600 years ago, but it had many features of modern cities.

It was laid out on a grid system with broad thoroughfares and straight, evenly spaced streets. There were open drainage ditches, drawing excess water off the streets, public toilets, underground sewage lines, and they even had manholes, in order to service the sewers. The buildings were constructed of fired bricks. The majority appear to have had private toilets and showers, draining into the sewage system. There appears to have been public bathing facilities, granaries, assembly halls and garrisons. Then in about 1,800 BC, the city was mysteriously abandoned, perhaps destroyed by flood or some other disaster.[19]

The biggest mystery of the Harappan people was their history. It appears that the oldest cities, such as Mohenjo-daro, are among the most advanced. But where did this culture come from? It seems odd that the very first settlements that these people ever built would have the most advanced, well-developed features. Normally we would expect to see a growth curve, with gradual improvement in architecture and city planning. Now Badrinaryan claims to have found the roots of the Harappan culture.

Why has it been missing for so long? According to dates of the Khambhat artifacts combined with Badrinaryan's geological calculations, the first city was destroyed in an earthquake, and subsequently sunk under the sea in about 7600 BC, nearly 10,000 years ago. The second city was built on higher ground. That city too, according to Badrinaryan, was "affected by faulting due to earthquakes around 4000 BCE, and was destroyed and submerged by an earthquake in 2780 BCE ($\pm$ 150 years),"[20] a little over a hundred years before the founding of Mohenjo-daro. This would explain why Mohenjo-daro was so advanced and where the people who created it came from.

This gives us evidence of yet another culture, even older than the Minoans, which also may have been destroyed by a natural disaster and, like the Minoans, saw their great and advanced cities sink into the sea. Badrinaryan goes on to connect these cities with the Kachchi folk songs, and the myths of the fabled city of Dwarka, home of Lord Krishna of the Mahabharata.

Some Disclaimers

Not unlike a three-word review of the Bible, this find is "important if true." Unfortunately almost nothing is known about Mr. Badrinaryan or these ancient twin-cities, outside of the article mentioned above. Author, Graham Hancock, reports having personally met him, and Badrinaryan's paper appears to have been well researched and authoritatively written. Still, these finds have yet to be corroborated by other archaeologists. And a quick review of this research suggests that some skeptics claim that the "cities" under the bay may be natural rock formations, and that the artifacts may be real artifacts that have been washed into the bay from other sources, due to storms and ocean currents.[21] That doesn't mean that they are not Harappan artifacts; they look as though they are. And it is also well established that the bottom of the Bay of Khambhat was once dry land thousands of years ago, when sea levels were much lower. So, with or without the twin cities under the bay, evidence of the long back-history of the Harappan people may indeed have been found in the bay of Khambhat. But, because of these other issues, this research is even more tentative than the previously mentioned Atlantis research.

Just like the Minoans, another problem with Khambhat is the timing and cause of their demise. If the geological history presented above is accurate, then it looks like this was not due to crustal displacement, but rather simple plate tectonics, though it may have been influenced by a large heavenly body, there is certainly no evidence of that. Also, the carbon dating of the artifacts suggests no disaster occured 24,000 years ago (one precessional cycle) but rather it was flooding 5000 years ago that sunk them.

The closest thing we have to a match with our myths and scientific theories is a pattern of intense earthquakes from 4,000 BC to 2,780 (± 150 years), eventually sinking this portion of the land, beneath the bay. And, these dates do straddle the date of 3,114 BC, which is the date of the beginning of the Mayan Long Count. This provides a somewhat plausible scenario that the events in India and Mesoamerica may have both been part of a global phenomenon of increased earthquake activity, floods, and perhaps also increased volcanic activity, but it is unclear what the cause of this might have been. If this instability lasted from 4,000 BC to as recently as 2,780 BC, then 3,114 BC is as good a date as any to mark the beginning of a new age. And 3,114 BC might just mark the exact date of an earthquake or flood that occurred in Mesoamerica at that time.

Assuming that the above research indicates tentative proof of the many antediluvian myths that abound around the globe, this still does not fit important criteria that we are looking for. First, neither the Minoans or Atlantis nor any Harappan city seems to be related to the precessional cycle of 24,000 years. And secondly, these appear to be somewhat localized events, and don't fit any pattern of global destruction. Having said that, there may be a global cataclysm related to the finds at Khambhat. What if the sinking of the last city of Khambhat and subsequent founding of inland cities such as Mohenjo-daro was part of a global pattern at that time? That might fit our criterion.

A Cataclysm in the 4th Millennium BC?

Any historian or archaeologist will tell you that recorded history
began roughly 5,000 years ago, at just about the start of the Mayan
Long Count in 3,114 BC.[22] There are two odd features of this.
First, civilization seems to have broken out everywhere at once.
Fairly advanced civilizations appeared in Egypt, Mesopotamia,
Mesoamerica, South America, China, the Mediterranean, Great
Britain, and the Indus Valley, all about the same time. There were
massive building projects all around the globe at that time as well,
such as Newgrange in Ireland, Stonehenge in England, Skara Brae
in Scotland, Hagar Qim in Malta, the Hypogeum of Hal-Saflieni
in Malta, the White Temple and Anu Ziggurat in Uruk, Shahr-i-
Sokhta in Persia, Mohenjo-daro in the Indus Valley, and the Great
Pyramids of Egypt, and all of them date from roughly the same
period. And civilizations with rich traditions of science, literature,
mythology, art and architecture were popping up everywhere. The
Sumerian civilization, the Yang Shao civilization of China, and the
first dynasty of Egypt, and the Minoans, all date to about the 4th
Millennium BC.[23]

Secondly, as already mentioned, they were all fairly
advanced compared with the civilizations that followed. That seems
strange. We all know that cultures are supposed to progress and
develop over time, not digress, and become less advanced over
time. Similar to the Harappan civilization, the oldest structures in
Egypt, the Great Pyramids, are the most advanced in the history of
Egypt. And of course many of these structures such as Stonehenge,
Newgrange, or Hagar Qim are marvels of engineering that were
not surpassed for thousands of years. One could argue that the

architecture of the Egyptian Pyramids have never been surpassed. In every case, these building projects occurred in about the same time period and what came later was always less advanced.

We see this same pattern with the stone masonry in Central and South America. Many believe that the massive and advanced stone work of Tiahuanaco, Machu Piccu, and the giant Olmec carvings are older than archaeologists think, and may well date to the second or third millennium BC.[24] The Olmecs, who predate the Maya and Aztecs, were the ones responsible for creating the brilliant Maya calendar system, and also created some huge, very skillful, anatomically correct human sculptures. When you compare the massive stonework mentioned above to that of the Maya or Aztecs, which followed, the earlier civilizations appear to be more advanced technologically and artistically.

The same is true in Mesopotamia. The Assyrian King Aššurbāniapl collected ancient clay tablets from far and wide, forming a library in Nineveh in the 7th century BC.[25] He apparently collected tablets of ancient knowledge including literature and astronomical information that were possibly thousands of years old at the time. Yet, the culture that he lived in was not believed to be as sophisticated or erudite as the works that he sought to preserve. It appears that with each new millennium in the Mesopotamian region, the culture became gradually less sophisticated and less knowledgeable. Eventually Nineveh was sacked and burned by the Babylonians, and the region descended even further from its former glory. From Sumer to the Assyrian culture, to the chaos that followed, each generation appears to have lost some of the knowledge and culture of their ancestors. How is that possible? That's not how society is supposed to progress.

The Turning Point

Something happened in the 4th Millennium BC, but what.
One possible scenario that fits the archaeological data might go
something like this. A great and advanced global civilization was all
but wiped out by some global catastrophe, possibly the great flood
that is often mentioned in ancient mythology. The survivors might
have migrated and created a new civilization, which started out as
very advanced, then gradually they forgot the advanced knowledge
and technology of their ancestors. When did this occur? Well, there
is no record of any advanced civilization anywhere much prior to
about 3,100 BC. And what cultures did exist, such as the Varna
and Çatalhöyük cultures disappeared at right about this same time
period. This was the start of what we call the Bronze Age. Prior to
this date, it is believed that humanity was coming out of the Stone
Age. In Egypt, Sumer, Crete, Shahr-i-Sokhta, Mohenjo-daro, and
elsewhere, the Bronze Age broke out at about the same time all
over the world. The Bronze Age of China soon followed, but as I
documented in my previous book, *The Secret Tao*, in many respects
the early Bronze Age dynasties of China were more barbaric and
less sophisticated than their supposedly stone age ancestors, the
Yang Shao people of 3,100 BC.[26] And we've already documented
the Old Copper Culture of North America, which appeared soon
after.

So roughly 5,100 years ago advanced civilizations
developed all over the world at once. And, at about the same time,
they all started to gradually de-evolve, losing wisdom, and losing
ancient knowledge. They lost the building techniques that we see

in the Egyptian Pyramids or Stonehenge. They lost the knowledge of the electrical devices found in Akrotiri, the knowledge of how to prepare maps with proper longitude and latitude, as demonstrated by Charles Hapgood.[27] They seem to have eventually lost knowledge of advanced city planning, knowledge of the precession, knowledge of the earth orbiting the Sun, knowledge of binary star systems, and much, much more. Some things, such as warfare, greed, and classist societies did increase,[28] but it seems that the noblest aspects of humanity, especially scientific knowledge, were gradually becoming lost, only to be rediscovered thousands of years later.

So what happened about 3100 years ago? Well, according to many geologists it may indeed have been a time of global geothermal instability and increased seismic and volcanic activity. Additionally, some spectacular, if not particularly sudden, flooding occurred. In addition to the possible Harappan cities in the bay of Khambhat, the history of Mesopotamia may also lie under the sea, and might have occurred about the same time.

Jeffery Rose, a research fellow at the Institute of Archaeology and Antiquity, at the University of Birmingham has been doing some very interesting underwater archaeology in the Persian Gulf. In a recent article he cites the fact that the Persian Gulf was a lush green fertile plain, about the size of Great Britain, prior to the melting of the last ice age.[29] He calls this land the Arabo-Persian Gulf Oasis, as it was between the deserts of Arabia and Persia, and it was a lush flood plane at the bottom of a confluence of four major rivers, two of which are the famous Tigris and Euphrates rivers. So this "oasis" may well have contained the site of the mythical garden of Eden. And, just as in the bible, the people there were progressively driven out, and eventually their land was completely flooded, as in the epic of Gilgamesh.

After the ice age ended and the ice began to melt, this large basin of land began to fill with water. Slowly over thousands of years, somewhere between 6,000 BC and 4,000 BC, according to Rose, the area became the sea that we now know as the Persian Gulf.[30] This explains the sudden appearance of cultures all around the Persian Gulf, dating to just before 3,100 BC, including Mesopotamia, ancient Iran, and the Indus valley. Recent archaeological digs in Iran, just north of the Persian Gulf have revealed that a number of sizeable and fairly advanced cities were established at about the same time, 3,200 BC; these include Shahr-i-Sokhta, Shahdad, Jiroft, and Tepe Yahya.[31] Note that these cities were about halfway between the contemporary civilizations of Sumer in the west and Mohenjo-daro in the east, and were located just north of what was the Arabo-Persian Gulf Oasis.

Such flooding may also explain migration into parts of what is now Arabia, Egypt and Yemen. And of course all of these cultures appear fully formed, with little history of progressive development, because they had probably been developing for 10,000 years in the now flooded Persian Gulf Oasis. All traces of their early attempts of civilization, like their Harappan neighbors, might now be completely lost to the sea.

Mesoamerica Prior to 3,100 BC

We know that sea levels rose all over the world at this time. Though we have less clear evidence in other parts of the world, there is no doubt that other early civilizations emerged about the same time in

the Americas, India, and China. For the purposes of validating the
Mayan mythology we need only remember that the great Olmec
civilization, beginning with the "Pre-Olmec" culture, began at least
4,500 years ago, within a few hundred years of the beginning of
the Long Count, 3,114 BC, a calendar system which was actually
created by the Olmecs. As mentioned before, the Olmec culture
appears very advanced. But where did the Olmecs come from.
Again, the oldest artifacts of the Olmecs may also be underwater.

There are vast areas off the coast of the Gulf of Mexico that
have not been fully explored. It is estimated that both the Yucatan
peninsula and the state of Florida had a land-mass that was more
than twice the size of today, and there was a strip of dry land nearly
as wide as Florida that extended all the way around the Gulf, which
is now submerged.[32] Though no Pre-Olmec sites have been found
in these areas yet, there are thousands of square miles to cover, and
no systematic effort has yet been attempted. Some archaeological
sites have reportedly been found in shallow water off the coast of
Cuba, under the Yucatan Channel, and underwater in Belize.[33] And
of course there is the now famous Bimini road in the Bahamas, off
the coast of Florida. This may just be a strange rock formation, but
appears to be a series of square blocks laid down, in a straight row,
three abreast like a road, about a half a mile long.[34]

Archaeological finds from prior to the third millennium
BC have turned up, but not where you'd normally expect to find
them. In the last decade underwater archaeological sites have been
discovered in the Yucatan Peninsula in cenotes, which are caves that
are now filled with water, but were once dry thousands of years ago.
Reportedly a whole host of artifacts have been found in the cenotes,
which indicate evidence of human habitation going back 10,000

years or more. [35] Some people report finding roads, walls and other stone structures as well. This corresponds very well to the accounts of the Xibalba (underworld) in the Popol Vuh. In this way we can validate that there was indeed a time, prior to the recorded history of the Maya, in which the people of Mesoamerica lived in a now flooded underworld.

In South America, there are the ruins of Tiahuanaco a vast pre-Incan city with a population of more than 40,000 people. The city appears to have been occupied for a very long time, up until about 1,000 years ago. While most artifacts at Tiahuanaco are more recent, 2,000 years old or less, there is one artifact that has caused quite a stir. In one unexplored area near the ruins, a local worker in the 1950's found the Fuenta Magna, a bowl with what appears to be Sumerian cuneiform writing on it, it has been dated to 3,100 – 3,500 BC.[36] Of course, skeptics claim it is a fake. Later, however, archaeologists found yet another artifact in the same area, a sculpture more consistent with the look of other sculpture at Tiahuanaco, and it too has cuneiform writing on it. We may never know when Tiahuanaco was actually founded because we cannot date the rocks using carbon dating techniques, we can only date when the rock itself was made not when it was carved. But if it was founded around 3,100 BC, as the date of the Fuenta Magna suggests, then it certainly fits with other fairly advanced civilizations springing up all over the world at the same time. And it even suggests a possible connection with the flooded ancient people of the Arabo-Persian Gulf Oasis, the Sumerians.

Global Flooding

In Asia, there are vast areas off the coast of China in the yellow sea near the mouth of the Yellow river valley, extending all the way out to Korea and the southern coast of Japan, which were once dry land and are now submerged beneath the sea. This region is another area where some think that they have found a giant underwater stone structure, the Yonaguni monument.[37] Though, it is still unclear if it is human-made or merely a natural stone monument that was used by humans, but there is no doubt that this monument was on dry land, thousands of years in the past.

The flooding is not in dispute. It is a geological fact that there was global flooding at the end of the last ice age, finally concluding around 5000 to 6000 years ago, resulting in our present shorelines. The question is whether this explains the many flood myths found from around the world, and also explains the beginning of the current Long Count as being a little over 5000 years ago. This would indeed be seen as a new epoch of humanity as hundreds of thousands of people, possibly more, were forced to migrate to higher and drier land just prior to about 3,100 BC. This certainly seems to validate the Gilgamesh/Noah narrative that is so ubiquitous in myths from around the world. It also validates the great flood reported in both the Maya and Aztec mythologies. But there was another, even greater cataclysm that happened even closer to the end of the last ice age.

The Cataclysm of 9,500 BC

There is a wealth of data from all over the world that something horrific happened at the end of the Pleistocene, which officially ended around 11,700, years ago, plus or minus a few hundred years. This was the last major ELE, extinction level event.[38] This event involved the simultaneous extinctions of Mammoth in Siberia, Mammoth, Mastodon, Saber-tooth Cats, the Giant Beaver, Short-faced Skunk, Horses and Tapirs in North America, and the Giant Sloth, Cuverionius, and Toxodon in South America. Basically any animal that was over about 100 pounds at that time probably went extinct, all at once.

How did this happen? Well, we know that the climate changed radically at this time. There appeared to be a giant ice cap centered over the Hudson Bay and extended across much of North America up until about 18,000 years ago. Then that began to change. By 9,500 BC the climate had completely changed and the ice cap that covered northern North America was completely gone. [39] As Hapgood documented so well, at the same time it got suddenly much colder in the Arctic Sea and Siberia. Very suddenly! Hapgood cites evidence that Mammoth were frozen so quickly that the meat was still good enough to eat over 11,000 years later. He also cites evidence of Saber-tooth Cats, and fruit trees that grew 90 feet high in the Siberian islands approximately 12,000 years ago. That land is now a desiccated, barren wasteland of snow and ice.[40]

The suddenness of this change is also documented in the mass burial pits of animals of all sorts that died at the same time. Evidence of this kind was found in North America, South America, Siberia, Alaska, Europe and elsewhere. Hapgood quotes Professor Frank C. Hibben from the book *The Last Americans*, who wrote "The Pleistocene period ended in death. This is no ordinary extinction of a vague geological period which fizzled to and uncertain end. This death was catastrophic and all-inclusive. … The large animals that had given their name to the period became extinct. Their death marked the end of an era."[41] Hibben is quoted as going on to say "where we can study these animals in some detail, such as in the great bone pits of Nebraska, we find literally thousands of these remains together. The young lie with the old, foal with dam and calf with cow. Whole herds of animals were apparently killed together, overcome by some common power."[42] And the same pattern of collective animal remains have been found in bone pits in Germany, Siberia, and South America as well.

The End of an Age of Humanity?

There is no doubt that this was a catastrophic event, but was it the end of an age for humanity as well? The question remains as to whether this was the cataclysm that was documented in the ancient myths. Unfortunately we do not have evidence of civilizations being destroyed. This is because, as far as we know, there were no civilizations to be destroyed at that time. But that may not be completely true.

Just last year, archaeological digs at site 075 at Wadi Faynan, in Jordan, have revealed a major architectural structure, 4500 square feet in size.[43] Most astonishing is that the site dates to 11,700 years ago. It was perhaps a meeting hall and/or a food processing plant, possibly for the production of barley or pistachios. This completely changes how we perceive early human settlements, and it is not the only such structure. Not far away is the tower of Jericho in the Israeli occupied West Bank. This was a structure that was 28 feet tall, and 30 feet across at its base. It is estimated to be 11,000 years old, and it is believed that it might have been used for astronomical purposes, given its alignment with the solstice Sun.[44]

Perhaps the best evidence of human civilization existing 12,000 years ago can be found in ancient Anatolia. Excavations at

Göbekli Tepe have uncovered a massive temple complex covering 90,000 square meters, with 17 separate structures, and with carved stone pillars up to 18 feet in height and weighing up to 16 tons. The site is right where you'd expect to find it, in the northern part of the famous Fertile Crescent, the so-called birthplace of civilization. The most amazing thing about this site is its age. Archaeologists have dated the site to 12,000 years ago.[45] This makes it the oldest temple in the world, certainly the oldest building of its kind in the world. Archaeologists have further connected this temple complex to other even more extensive settlements nearby that came later.

A little further away and later by a couple thousand years we have the settlement of Çatalhöyük, which began around 9,500 years ago.[46] Archaeologists have reason to believe that this was likely a continuation of the people and culture that built Göbekli Tepe.[47] As mentioned earlier, there is now a direct DNA link between the people of Çatalhöyük and the Minoans that followed, who traded extensively with Egypt, Greece, and others civilizations at that time.[48] This shows that there is a thread of culture stretching from the earliest known civilizations in Mesopotamia and the Mediterranean, all the way back to the actual people who witnessed the cataclysm of 11,700 years ago, from their temple complex in Göbekli Tepe. So, there could well be a connection between the cataclysmic end of the Pleistocene and the stories of extreme cataclysms found in mythology. And, according to Plato, the catastrophe that destroyed Atlantis occured 11,500 years ago.

Questionable Finds at Tiahuanaco

The finds at Wadi Faynan, the tower of Jericho, and Göbekli Tepe raise some important questions for archaeologists. Could there have been other such civilizations from the same time period? If so, where might they be? It would be especially interesting to find such ancient sites in the Americas. Interestingly, there is possible evidence of such civilizations having existed.

In Hapgood's research, he has documented the violent geological history of the Altiplano lying along the Peruvian-Bolivian boarder. This is a plain at an altitude of over 12,000 feet, that contains Lake Titicaca and the ancient ruins of Tiahuanaco. Again, no one knows exactly who created Tiahuanaco or when, as the carving of rock cannot be carbon dated, and most of the artifacts are more recent. But the geology is another matter. The entire plain is tilted, and there is clear evidence that the lake was once much larger and closer to Tiahuanaco, but it drained away as the entire Altiplano area was driven up violently thousands of feet, in seismic activity; according to Hapgood, this was approximately 12,000 years ago, probably as a result of the earth crust displacement.[49]

The interesting thing about this geologic activity is that archaeological studies of Tiahuanaco indicate that it was a harbor town, on the shores of the lake.[50] But as far as geologists can tell the lake has not been near the town for many thousands of years, possibly as long as 11,500 to 12,000 years ago. This is because the movement of the lake was believed to be the result of the same violent geological upheaval that raised the plateau. This would date the founding of Tiahuanaco to at least 12,000 years ago. Could the ruins at Tiahuanaco really be that old? This means that this site might have been inhabited long before the fourth millennium BC, as mentioned earlier. Just a few years ago, most archaeologists would scoff at the idea. But we now know that there were other civilizations building monumental architecture at that time. If

Tiahuanaco were 12,000 years old, that would put it in the same time period as Göbekli Tepe, site 075 at Wadi Faynan, and the tower of Jericho. So, it may not be as far fetched as previously thought.

Graham Hancock, in his book *Fingerprints of the Gods*, has revealed that there are stone carvings in Tiahuanaco, which represent animals that went extinct at the end of the Pleistocene.[51] These include the Cuverionius, a proboscid relative to the modern elephant, and a Toxodon. If the inhabitants of Tiahuanaco carved these in the first or even the second millennium BC, as we believe, then how were they able to depict animals that had been extinct for more than 8,000 years?

Charles Hapgood cites numerous geologists who verify that the Altiplano was suddenly and violently driven up by a cataclysmic earthquake of unthinkable proportions as a result of crustal displacement. And Hapgood also provides archaeological evidence as well. He cites Dr. Ellsworth Huntington who completed an aerial survey of the arid and desert regions of Peru. Huntington noted that there are a number of "old ruins" and "an almost incredible number" of man-made, "terraces for cultivation," surrounding the Altiplano region, that are now "absolutely desiccated" and in many cases above the snow line.[52] This indicates that there was a thriving civilization here before the dramatic seismic activity forced the land up thousands of feet, to the altitude it is now.

Hancock also cites Professor Arthur Posnansky, a German-Bolivian scholar who has investigated Tiahuanaco for more than 50 years and knows the site better than anyone. According to Posnansky, there was a terrible flood here, which destroyed Tiahuanaco. He believes that the intense seismic activity heaved the lake up onto the shores of Tiahuanaco and probably also caused lakes from higher in the mountains to break their bulwarks and flood the Altiplano. He quotes Posnansky as saying "The discovery of lacustrine flora, Paludestrina culminea, and Paludestrina andecola, Ancylus titicacensis, Planorbis titicacensis, etc., mixed in the alluvia with the skeletons of human beings who perished in the cataclysm … and the discovery of various skeletons of Orestias, fish of the family of the present bogas, in the same alluvia which contain the human remains"[53] proves that it was flood that destroyed the city. He goes on to quote Posnansky as saying that the "chaotic disorder among wrought stones, utensils, tools and an endless variety of other things. All of this has been moved, broken and accumulated in a confused heap. …Layers of alluvium cover the whole field of the ruins and lacustrine sand mixed with shells from Titicaca, decomposed feldspar and volcanic ashes have accumulated."[54]

Finally, there are the strange rock formations of Markawasi. This site in Peru, in the same part of the world as Tiahuanaco, and even closer to Machu Picchu, is known for what looks like amazingly large stone sculptures. There are what appears to be carvings of all sorts of animals and people too. Skeptics believe that they are just unusual rock formations that look like sculptures, but clearly some of the rocks have been carved to some extent. Of course it could be a coincidence that one rock looks like a person or an animal. Even two or three such rocks could be a coincidence, but there are no less than 38 such formations at Markawasi.[55] The problem in determining whether they were carved or natural formations is two fold. First, if they were carved, the artist clearly used the natural shape of the rock, and merely embellished it to bring out the natural form of an animal or person, with minimal carving. Second, if they were carved, they are so heavily weathered that it is very hard to make out what appears to be definite but faint forms. If carved, they must be many thousands of years old.

One of the most impressive features of the Markawasi site is that, not unlike the stone carvings at Tiahuanaco, Markawasi contains images of South American animals that went extinct at the end of the Pleistocene.[56] This again indicates that there might well have been a culture of people who witnessed this cataclysm and have preserved it in myths, which persisted up until the time of the Incas, and further north in the Olmec culture, and the Mayans who followed them.

The Smoking Gun

This extinction level event, which ended the Pleistocene Age, and occurred approximately somewhere between 11,500 to 12,000 years ago, is the smoking gun. This truly is a global event, and an incredibly destructive event. It appears to have involved major climate changes around the planet at the end of the last ice age, the death of millions of animals, and extinction of hundreds of species.[57] In addition to cataclysmic flooding, it involved considerable geological, volcanic and earthquake activity as well. This is exactly the kind of evidence we have been looking for.

Additional climate evidence supports the idea that the poles moved to their current position at about that time (9,500-10,000 BC), and will be presented in a later chapter with the rest of the evidence of crustal displacement. But suffice it to say that there is much more evidence than what is presented here, indicating that this was a major, global cataclysm, consistent with

the most dire myths of past cataclysms.[58] It is revealing that it is
not just mythology that claims this was the end of an age, and the
beginning of a new age for the world. Mainstream science marks
this point in history as the end of the Pleistocene Age and the
beginning of the Holocene Age. This is because of the dramatic
change in climate, massive extinctions, and subsequent change in
sea level and coastlines of the world, it was truly a new world that
dawned at this time.

Even archaeologists mark this same time as the beginning
of what they call the Neolithic Revolution, which was marked
by the sudden creation of buildings, temples, agricultures and so
on. So in many fields of mainstream science, including biology,
climatology, paleontology, geology, and archaeology, this was the
end of one age and the beginning of another.

Atlantis II

One of the most interesting things about this Pleistocene cataclysm
is that it also relates to the myth of Atlantis. Remember that,
according to Plato, the Egyptian priest met with Solon around 500
BC, and told him that Atlantis was destroyed 9,000 years earlier.
That would mean that Atlantis was destroyed around about 11,500
year ago. It is uncanny that Plato would just happen pick this
exact date, if he did not have evidence of a major cataclysm from
that exact time. Also he says that Atlantis lied beyond the Pillars of
Hercules, which would mean somewhere across the Atlantic Ocean.
Was he referring to Tiahuanaco? It was a major civilization might
have indeed existed in the Americas at that time, right where Plato
said it would be.

We now have not one, but two potential candidates for

the mythical lost continent of Atlantis, and in both cases the lost continent is actually America! Given all the details presented thus far, the case for the Minoan civilization is much stronger. Nonetheless, I don't believe that it's just a coincidence that Plato just happened to pick a date in history that really does coincide with a known major and global cataclysm. That's a bit of a stretch, even for a die-hard skeptic. It is more likely that the Egyptian priest probably got his cataclysms mixed up. As we have already shown, the Egyptian civilization appears to have begun not much before 3,100 BC, after the last great flood. So it is unlikely that the Priests of Egypt would have had an accurate history going back 9000 years.

The Egyptians would indeed have had information about the Minoans, their mysterious continent across the Atlantic, and their sudden demise, apparently sinking into the sea. And, through ancient myths and legends they might have also known that some great cataclysm happened 9,000 years in the past, especially if it involved crustal displacement. Clearly, no culture can completely erase the memory and subsequent legend of something like that. But it's doubtful that they had that much detail about any civilization that existed that far back. So it seems likely that the well documented civilization and ultimate destruction of the Minoans might have been confused with the most famous event in history, the day the Sun actually dawned in the wrong direction. After all, they probably knew that some great cataclysm happened 9,000 years earlier; they might have even known that a great civilization was destroyed beyond the Atlantic. It's reasonable to think that that famous event must have been when the Keftiu, as they called the Minoans, were destroyed.

One might argue that having two potential rivals for the

historical Atlantis undermines the proof of finding Atlantis. I disagree. What this shows is that there is actually ample proof of civilizations across the Atlantic being destroyed by flood. We can demonstrate this with a simple hypothetical example.

Imagine hundreds of years from now, in a post-apocalyptic future, people argue whether humans had ever visited the Moon. Imagine that there is a popular legend that people did make it to the Moon once, but they never returned, and no one knows why. Eventually, in the future, they too go to the moon. But when get there they find evidence of not one landing, but many such landings. Does that disprove that we ever went to the Moon? No, quite the opposite. Likewise, having multiple different civilizations in the Americas that might have been destroyed by flood hardly disproves the myth of Atlantis. It just shows how much history there was prior to the time of Plato, and that there may have been more than one civilization in the Americas destroyed by flood or disaster in ancient prehistory. There have actually been multiple cycles of creation and destruction in the distant past, just as the Mayan mythology had described. And there have been multiple epochs of human history, which until very recently we knew practically nothing about.

Evidence of Past Cataclysms Continues to Build

In this chapter we sought to collect evidence that a) humanity has been wiped out a number of times by different cataclysms, each time rebuilding and becoming quite advanced, and each time being destroyed again. And b) the specific apocalyptic features reported in the myths of the past do indeed reflect actual cataclysms that have occurred on earth.

We have found evidence of multiple events around the world, mostly flooding, in which advanced ancient civilizations were destroyed or at least forced to migrate, and in the process lost much of their culture and had to start all over again. This seems to have happened all over the world in about the same time period, about 5100 years ago, and again around 11,500 years ago, give or take a few hundred years. And the research documenting this is still coming in every day.

I cannot stress enough how overwhelming the recent evidence is on this topic in the archaeology literature. Every year archaeologists are finding and publishing more and more new finds of complex and fairly advanced civilizations that sprang up around the world in the third millennium BC. In the last issue of Archaeology magazine I found three different articles documenting new bronze age sites and artifacts, recently discovered. One showed evidence of stone buildings in the Central Asian Steppes, dated to 4,500 years ago, with the oldest domesticated grains ever found in the area.[59] Another article documented a cache of well-made bronze axe heads, in a building that was 35 feet wide and nearly 150 feet long near Dermsdorf Germany. This was one of a large number of buildings found at the site and only one of a number of sites in the area, belonging to the Únêtice culture, dating to approximately 4300 years ago.[60] And yet another article in the same issue documented evidence of fairly sophisticated ancient board games, not unlike our monopoly, found in Chiapas Mexico and dated to over 5,000 years ago.[61] This was supposedly before the beginning of the Pre-Olmec culture.

And the evidence that these cultures sprang up all at

once because of migration due to flooding is also growing. In another recent issue of Archaeology magazine, there is an article documenting the prehistory pearl trade and pearl jewelry industry around the Persian Gulf. Archaeologists have found evidence of elaborate pearl jewelry in graves and tombs dated up to 7,000 years ago, all along the Persian Gulf. [62] What's so interesting is that these finds are not just in the fertile crescent of ancient Babylonia and Sumer where you would expect to find them, but they are found all along the entire stretch of the Persian Gulf, from modern day Kuwait, south to Bahrain, Quatar, in the United Arab Emirates, and in Oman. And furthermore, some of these finds predate the Sumerian civilization by nearly 2000 years. This clearly indicates evidence of another, yet older culture, centrally located between these points. And, 7000 years ago, that would be in the lush fertile valley that is now at the bottom of the Persian Gulf.

As I have already documented, we have also found evidence of an even larger, even more cataclysmic extinction level event, which would have likely created many of the exact apocalyptic features reported in the mythology, dated to the time of ancient Atlantis about 11,500 to 12,000 years ago. So, we have definitely validated that the myths of past cataclysms could have been based in fact. Given the above research, I'd now say that it would be very hard to believe that they were not based in fact.

Given this conclusion, we can see why our ancestors might have thought that these events could happen again in the future. So, at least one piece of the puzzle is in place. It is validating to know that the ancient myths were not pure fiction. If they did not record actual past cataclysms, then it is an incredible coincidence that their stories just happen to be identical to actual events that real people experienced so many thousands of years ago.

11

EVIDENCE OF EARTH CRUST DISPLACEMENT

One of the most debated scientific theories of apocalyptic doom has to be Charles Hapgood's theory of crustal displacement, explored at length in Chapter 6. What follows is the research evidence presented by Hapgood, along with outside research both corroborating and contradicting his work.

As you will remember Hapgood was a historian who, while examining ancient maps, discovered that renaissance mapmakers, using ancient source maps, had included things that made no sense, given what we know of history. His research indicated that thousands of years ago some sophisticated cartographers had mapped Antarctica, used knowledge of both latitude and longitude, and showed the South Pole in a different location than it is in now.[1] These were all puzzling anachronisms and mysteries that existing science could not solve. Subsequent to this he and a geologist named James H. Campbell came up with their theory of earth crust displacement.[2]

The theory of earth crust displacement states that that every 20,000 to 30,000 years, the entire lithosphere (outer crust) of the earth slips around the mantle underneath it. This can cause a difference in latitude of up to 30° in the physical location of the poles. This highly controversial theory explained a number of observed yet unexplained phenomena. It explained why we have periodic catastrophic events such as evidenced at the end of the Pleistocene. It also explains the coming and going of the ice ages, also evidenced at the end of the Pleistocene. Finally, it explained what appeared to be a shifting of the physical South Pole, as evidenced by the ancient maps he discovered.

Dr. Hapgood included a wealth of scientific data in his book, *Path of the Pole*. In fact, well over 300 pages out of the more than 400 pages of the book is dedicated to either presenting scientific research or presenting documents and references, which support that research. There is no way I can cite all the evidence presented in his book, and discuss objections to it, and discuss conclusions that we can draw from all this research, all in one chapter. In a sense, that is true for this entire book. This book is like the reader's digest version of the ancient apocalyptic prophecies. To really do all of the source documents justice, I'd have to produce an encyclopedic series of volumes, with a dozen books or more, consisting of nothing but translations of ancient texts, scientific theories, and reams and reams of scientific evidence. So what follows is only a summary of Hapgood's research. To really appreciate his work, I definitely recommend reading his book, *Path of the Pole*, and just for a little more background you may want to also read his book, *Maps of Ancient Sea Kings*, and Graham Hancock's *Fingerprints of the Gods*.

Hapgood's Research Evidence in Three Parts

So as to not reinvent the wheel, in a manner of speaking, I will forgo with all the evidence supporting the conclusion that the end of the Pleistocene Age was a fairly abrupt extinction level event, involving a major climate change and the end of the Ice

Age. At least that much is not disputed among scientists. Those points are conceded by mainstream science in the fields of geology, atmospheric science, biology, and paleontology. This is also true for evidence of the rise in sea levels after the end of the Ice Age. So the only sticking point is what happened that could have caused this event. And that is still a debatable point, and one in which Hapgood's crustal displacement theory looms large.

Hapgood presents three type of evidence in his book. First is the geomagnetic evidence. He presents a wealth of data regarding the orientation of iron particles in rocks. By dating when the rock was formed, and looking at the orientation of the iron particles within it, we can get a picture of where the north pole was when that rock was formed. By then looking at rock samples over millions of years we can see if there were pole shifts and where they were.

Secondly, Hapgood presents the very same evidence that prompted his theory in the first place. Namely, that during the last ice age there was what you might call a polar ice cap over North America, while Siberia, parts of the Arctic Sea, and parts of Antarctica were enjoying an unusually warm climate. He documents this climatological evidence as well or better than any other evidence. There is no doubt about this fact either. Mainstream science would stipulate this point as well, except that they cannot completely explain it, so it remains a bit of a mystery.

Thirdly, he presents those ancient maps, which is how he began his investigation in the first place. Many people focus on the historical and archaeological significance of the maps. It seems incredible that some civilization existed up to 12,000 years ago that had sophisticated map-making skills, which we have no knowledge of today. But with regard to earth crust displacement theory, that is not the interesting part. The really important part is that these maps document what the world looked like at the end of the last ice age, and astonishingly it is quite accurate. That is, we actually have a map that shows the ice-caps exactly where we now know they were, based on the climatological research.

There is one more piece of evidence that I find compelling. That is the cross-correlations between these different areas of research. Imagine if the geomagnetic properties of rocks indicated that the north pole used to be in Hudson Bay during the last ice age, but the climatological research reveals that the polar ice cap was over Siberia. That would pretty much sink his theory. But that's not what we see. Quite the opposite. Every bit of evidence collected tends to corroborate every other bit of information. And the corroboration of these checks and balances is itself important evidence. That is a very brief summary, now let's dig into his research a little deeper.

Geomagnetic Research

A major component of Hapgood's research is geomagnetic research. Hapgood explains, "iron is a magnetic substance; that is, a substance which will itself become magnetic when exposed to a magnetic field, and will therefore align itself with the lines of force of the earth's field."[3] He goes on to explain that when rock is formed, such as in volcanoes, the tiny iron particles in these rocks become tiny compasses, lined up with the earth's magnetic field. But he also notes that these tiny particles actually give more information than a mariner's compass because they indicate both the direction and distance to the pole. Unlike a compass needle, the particles tilt down, proportional to the distance from the pole. Right over the magnetic north pole, the particle would be absolutely vertical So, he explains, "The horizontal angle is called the variation; the dip is called inclination. The variation give the longitude of the sample relative to the present magnetic pole, and the inclination gives the latitude."[4] And of course magnetic rocks align themselves to the earth's magnetic field at the time they were formed.

Hapgood acknowledges the many limitations and problems with this line of research. Local movements in the crust could move the rock hundreds of feet along a major fracture. But this can be accounted for by basing estimates on many rock samples from a wide area, not just one or two samples from one area. "Another problem arises from the fact that the magnetic field of the earth does not stay put," Hapgood wrote. But he cites that "the average position of the magnetic pole over the whole period will coincide with the earth's axis of rotation."[5] So he says that it's simply a matter of taking samples from a rock thick enough to represent the sedimentation of several thousand years. "Or if the samples are from lava flows, they have to be taken from successive lava flows indicated as having occurred fairly close together in time."[6]

He goes on to write that if the sample was deep in the earth originally, it may have been subjected to heat and pressure sufficient to destroy its magnetization or alter the direction of the field. But, again he points out that this problem "may be eliminated by the simple means of basing estimates of the position of the pole at a particular time on many samples taken from different places far apart in different countries or continents."[7] He suggests that when "many samples are assembled, and the results are averaged to eliminate the errors due to local factors, we begin to have a fairly reliable indication of the position of the pole at the time when the rocks were laid down."[8] He concludes his discussion of methodology by saying that "a vast amount of research has been done, and the state of the science had advanced until we can say that the present findings of pole positions are reasonably reliable."[9] And he estimates his findings to be 95% accurate. In science we say that this research has a statistical p value =.05. This is the minimal statistical threshold to establish statistical significance. This confirms these findings would appear by random chance in only 5 out of 100 times.

There is no dispute that the location of the pole wanders relative to the earth's crust, but this is believed to occur over the course of millions of years. Hapgood's research challenged this view. In the words of Hapgood "magnetic evidence in contradiction of this view, however, now exists."[10] He cites evidence from volcanic deposits in Japan by Nagata, Akimoto, and others, and from Soviet studies as well. He found a total of 229 specific positions for the north pole over eons, but acknowledges that many of these may be duplicates from different continents and may end up locating the north pole in the same place at the same time. His research plotted pole positions over the last 100 million years, and documented movement as recently as approximately12,000 years ago, plus or minus a few hundred years. He found a number of decisive shifts in the north pole, and two complete magnetic pole reversals.

His book presents page after page of detailed findings from various rock samples. The overall conclusions he draws are that in addition to what is often called "true polar wander" that all geologists concede has occurred over millions of years, he has found geomagnetic evidence in rock samples that the pole has moved at least three times in approximately the last 100,000 years. The summary of evidence suggests that somewhere around 75,000 years ago the pole shifted from somewhere in the Yukon territory to the Greenland Sea. Then around 50,000 years ago it shifted again to the Hudson Bay. This explains North America being covered in a mile thick ice sheet during the last ice age. Then, around 12,000 years ago, the end of the Pleistocene, the pole moved to its current location.

There are some interesting features in his conclusions that would argue against these findings being purely erroneous. First of all, he approximated estimates of pole positions from a number of samples, each from a number of different sites from different locations. Secondly, these results indicate a very predictable pattern of shifts, never more than 30° at a time, and always in exactly the same direction and angle. If he was just picking up noise, as we say, and these findings are not meaningful, then it would be an extreme coincidence that a consistent and predictable pattern of polar movement would appear out of the data set. Even the timing of the shifts is fairly consistent, occurring every 20,000 to 30,000 years, on average.

Summary of Geomagnetic Data

Based on the geomagnetic data alone, there appears to be compelling evidence for fairly rapid shifts in the earth's crust, moving the north pole thousands of miles, and that this has occurred several times in recent geologic history (the last 100,000 years). In the second edition to his book, Hapgood says that the data from geomagnetic rock samples indicates that this movement was not terribly rapid, that it might have taken 2,000 to 5,000 years to complete each shift. It is still not clear, however, whether that was a steady continuous movement, or a number of sudden slips, over a period of thousands of years. More recent research, as mentioned before, has suggested that the climate change at the end of the Pleistocene was very rapid, possibly a matter of days! But this could also have been due to the last of many little slips in the crust over a period of up to 5,000 years. The last of these shifts, approximately 12,000 years ago, could have been so destructive because of its speed of movement, not because of a particularly large distance of movement.

Climatological Evidence

When we talk about climatological evidence, we are really talking about exactly what areas of the world were cold during the last ice age, and which were warm. Where were the massive ice sheets located in the last ice age? And if they covered North America, then what was the climate at the poles like at that time? The idea is quite simple. If you could look at the climate patterns of the last ice age, you'd clearly see where the polar ice caps were, and where the equatorial regions were, ice age, or no ice age. If there was a global ice age, then the polar caps should reach down from the north pole all the way down into the lower latitudes. But, if there was a shift in the location of the poles, then we'd find polar ice caps centered at a geographic location that is now at a lower latitude, and no polar ice cap found within the arctic circle, where it should be.

So, what does Hapgood's climatological research reveal? Coincidentally, we find evidence of polar ice caps exactly in the same places and at the same times as the geomagnetic evidence suggests. Prior to 18,000 years ago, there was indeed a thick sheet of ice covering North America, centered over Hudson Bay. Strangely enough, at the same time temperatures were more and more temperate as you move further north. Parts of the Arctic Sea were much more temperate, and Siberia was enjoying a veritable heat wave. Then, between 12,000 to 18,000 years ago the location of the ice caps began to move.[11]

By far the strongest climatological data is for the last major climate change of 12,000 to 18,000 years ago. But he does report finding the same pattern around 50,000 years ago. Again, the climatological data suggests a shift in ice deposits from the Greenland Sea to the Hudson Bay, coincidentally right at the same time that Hapgood found evidence of a pole shift, and right in the same locations he predicted as well. Again, he found a similar pattern of evidence for 75,000 years ago, from the Yukon to the Greenland Sea. Some minor regional variations between the climatological data and the estimated location of the pole are discussed in the context of land masses vs. oceans. As oceans melt, they become dark blue, and actually soak up more heat than dry land. This adds a slight variability to the data set. And, as we have seen with the "little ice age" in Europe, between the 16th and 19th centuries, complex ocean currents that produce the thermohaline circulation in the Atlantic can have major climatological consequences, even when the pole location is constant.

Of course climatological data is not as simple as I have presented it here. There were a number of small periods of

glaciation, and interstadial periods during each of these larger
periods of time. But Hapgood and geologist James H. Campbell
provide a meaningful context that explains all of the varying
climatological data, from various parts of the earth. It accounts for
all the ebbs and flows of various periods of glaciation and so-called
"ice ages" with one elegant theory that covers a wealth of data that
was otherwise contradictory and confounding. I have, of course
simplified it, but the result is the same, only even more amazing.

Hapgood actually accounts for many more climatological
details than I have presented here, including the Cordilleran
Ice Sheet, the Early Würm Glaciation, the Brörup-Odderade
Interstadial, the Wisconsin Glaciation, the Late Würm Glaciation,
the Sangamon Interglacial, the Farmdale Advance, the Tazewell
Advance, the Brady Interstadial, the Bölling Interstadial, the
Alleröd Interstadial, just to name a few of the many climate changes
of the last 100,000 years. All of these changes, and many more, are
accounted for with his pole-shift hypothesis.[12]

Summary of Climatological Data

Hapgood's climatological research supports all of his estimates
of previous pole movement, which was based on geomagnetic
research. I would estimate the odds of this occurring by random
chance to be considerably less than 1 in 20. So, both of these
areas of evidence, geomagnetic and climatological, together
constitute sufficient checks and balances to support his thesis, with
a confidence level of 99% or better. This particular data has not
been contradicted in the last 40 years since he published his latest
findings. The only difference, as mentioned before, is that now it
appears that the climate shift of 12,000 years ago was more swift

than even he imagined. If it indeed took thousands of years for the pole to move 30°, then the last few days of this event must have constituted a significant percentage of that movement.

Archaeological Evidence

As mentioned at the outset, this all came about because of some puzzling maps that Dr. Hapgood found while doing historical research. Ironically, though these maps were the first evidence to support the idea of a polar shift, they are now the least compelling.

According to Hapgood's research, the ancient maps he discovered were each based upon older source maps. Some of these source maps have a provenance that appears to be many thousands of years old. How do we know? Because one map depicts the world as it looked in 4,000 BC, and another depicts the world as looked even longer back in time. Of course the most astonishing item he found was a map with parts of Antarctica, free of ice. The same polar ice cap that geologists thought had been there for millions of years.

Again, his climatological research validated his archaeological data because it turns out that there is evidence that parts of Antarctica were indeed ice-free until the end of the last ice age, and for sometime afterward.[13] This means that it is possible, you might even say probable, that someone did chart the world, including Antarctica, thousands of years ago. This does two things. Again, it lends support to his pole shift theory. In fact, one map of Antarctica, the Oronteus Finaeus map, actually pinpoints the South pole as being exactly where the geomagnetic data says it should have been during the last ice age.[14] But this research also lends supplemental support for the idea that a civilization like Atlantis really did once exist. In order for this map to exist, the civilization that produced it would have to be very advanced, and was probably destroyed at the end of the Pleistocene. According to Plato this civilization perished 11,500 years ago, and this is within a few hundred years of the date that scientists now say that a sudden climate change event occurred, killing millions of animals around the world and causing global mass extinctions.

If you ignore references to Atlantis, and just look at the location of the South Pole in the Oronteus Finaeus map, then the results are quite impressive. He is three for three. The geomagnetic evidence fits with the climatological evidence, both of which predict a shift in the earth crust and an estimated location for the South Pole during the last ice age, different from where it is now. And, lo and behold, the Oronteus Finaeus map pinpoints the South Pole right where the geomagnetic and climate research said it would be. Coincidence? If so, it's a pretty unlikely one.

The biggest problem with these maps is just accepting that any advanced civilization could have been around that long ago. That's why I suggest setting aside the comparison to Atlantis, because I feel that the geological evidence of these maps is actually stronger than the archaeological evidence. I don't know where the information to make these maps came from. They may have come from Aliens for all I know, but you can't get away from the fact that the map shows the South Pole right where Hapgood's geomagnetic and climatological data indicates it should have been 12,000 years ago.

Outside Research Evidence of Hapgood's Theory

One of the frustrating things you find as a researcher is that there are often holes in the scientific data. A lot of times, you plan on researching a topic and are expectantly waiting to see the results, only to find that no one has bothered to collect the data. Ironically this is one of those cases. Oh there's a lot of talk about Hapgood's theory in articles, reviews, and online sites. Much of it is hyper-skeptical, a lot of it is enthusiastically supportive, but most of it produces little or no evidence one way or another. I am not a geologist, but I think that even I could do a better job of evaluating Hapgood's theory than what I've been able to find in the research literature. What follows is a representative sample both supporting and refuting earth crust displacement theory.

The Einstein Connection

Ironically if you do an internet search for "Hapgood's Theory" or anything specifically related to "Hapgood" and focus your search just on scientific journals in the earth sciences, only two articles come up, and they are both on Albert Einstein. It's a testimony to the power of celebrity that while the field of geology generally ignores Hapgood's research, they do not fail to credit Einstein for his contribution to Hapgood's theory. In an article reviewing Einstein's contributions to earth science, authors Martinez-Frias, Hochberg, and Rull provide a fair and balanced review of Hapgood's theory, describing it as "very controversial."[15] Of course the focus is on Einstein's contribution, which was brilliant but ultimately flawed. As previously mentioned, Einstein agreed in principle with Hapgood's theory, but he was hung up on finding a force that is massive enough to trigger such phenomenon. So, the one piece of the theory that Einstein addressed was the potential cause of such a displacement of the lithosphere (earth's crust).

At that time it was believed that the polar ice cap, if unevenly distributed, could interact with the centrifugal forces involved to pull the earth's crust off of it's present position with regard to the poles. This was already developed with the help of geologist James Campbell, and Einstein clearly pondered this problem too. Alas, in spite of Einstein's help this is the least compelling part of the Hapgood-Campbell-Einstein theory, and it has been rejected as a theoretical explanation of crustal displacement. But, that still does not refute the evidence that Hapgood presented to validate that displacement did occur, even if we don't yet know how or why.

Theoretical Support for Planet Z

There have been very few papers published in scientific journals on the topic of earth crust displacement in the last decade, and the ones that have been published are surprisingly supportive. There was a paper written in 2004 as part of a series of papers from Cornell University researchers, which clearly supports Hapgood's theory concluding, "A scenario which causes such a rapid geographic polar shift is physically possible."[16] They do cite climate data as strong evidence to support a pole shift from Greenland to its present location. They put forward an alternative theory as to the cause of such a shift, and they suggest that a rogue planet with an eccentric orbit, which they call Z, could produce the kind of shift theorized by Hapgood, Campbell, and Einstein. The authors, Woelfli and Baltensperger, came up with a very interesting theoretical scenario, which actually helps to explain large global temperature variations during several million years.

This is very validating not only for earth crust displacement theory, but also for my own synthesis of the jovian mass theory combined with Hapgood's theory. The problem with this research is it is largely theoretical. They do validate the climate research which suggests that there was a "striking asymmetry of the ice cover during the Last Global Maximum."[17] But as I already said, the climate research is not in dispute; it is finding proof that displacement was the cause of that striking asymmetry that is needed. Still, it is interesting to see some theoretical validation for our hybrid catastrophe theory combining crust displacement and a rogue jovian mass.

More Mixed Support for Planet Z

Though not particularly recent, I found one article supporting earth crust displacement as a means of explaining complete geomagnetic reversals found in rock samples. An article by P. Warlow states that it is "shown that a wide variety of data is compatible with this hypothesis, not only from modern geological and related investigations, but also from astronomy and from ancient sources."[18] Warlow too advocates the same causal mechanism that I proposed in my hybrid theory, as did the above researchers, that a jovian mass passing by the earth could produce the necessary force to displace the earth's crust.

In a rebuttal to this article, a few years later in the same journal, V. J. Slabinski, writes that, after a complete analysis of the dynamical equations, such a crustal displacement "cannot be produced by the gravitational attraction of any known body in the solar system making a close passage past Earth."[18] Of course, neither Warlow nor I ever said it was a "known body." But Slabinski goes on to say that the inversion of the lithosphere "requires a 417 (times the) Earth-mass body passing at two Earth radii."[20]

The problems with these two articles are numerous. First, Warlow's theory is even more fantastic than Hapgood's, who says that the crust displacement is never more than 30°. Warlow proposes an entire 180° reversal of the poles. Secondly Slabinski claims to have completed an analysis of the "dynamical equations" but there is one important piece of information missing, preventing anyone from being able to analyze such equations, and that is the actual viscosity of the asthenosphere, which is the layer of mantle that the earth's crust rests upon. We can only make educated guesses what that might be based upon lava flows on the surface of the earth.

Support for Pole Wandering

There is no lack of support for the theory that the pole is physically moving. The only question is how far and how fast. It is an established fact the pole has migrated to its present position from other physical positions relative to the earth's axis; this is called *True Pole Wander.*[21] It is also clear that both the physical pole and the magnetic pole moves, independent of one another, and possibly for different reasons. What is in dispute is how fast this occurs. Most geologists believe that it takes millions of years for the pole to physically move, through plate tectonics and continental drift. And therein lies the most critical evidence against Hapgood's theory.

In an article in the Journal of Glaciology (1959), after
the first edition of Hapgood's book came out, R. F. King argues
against the geomagnetic data that Hapgood presents.[22] He says that
Hapgood misquoted the Nagata study on the geomagnetic evidence
found in a Quaternary volcanic region in Japan. He goes on to
suggest that other studies of magnetic rocks contradict Hapgood's
evidence, and that Hapgood's research "gives a rate of polar
movement of about one-third of a degree in a century, ten thousand
times faster than that indicated by rock magnetism."[23] But he does
not cite the source for any contradicting rock magnetism research,
nor does he say exactly how Hapgood misquoted Nagata, or what
was originally written. He then goes on to praise Hapgood's and
Campbell's research saying that "The great value of this book lies in
the author's collection of evidence in favour of polar shifts; whether
we believe in his particular conception of them or disbelieve in
them entirely, this evidence must be seriously considered."[24]

Dearth of Research

One of the biggest problems with this entire topic is the lack
of credible research directly supporting or refuting Hapgood's
findings. Oh, there are articles and blogs o' plenty on the topic,
but that's not the same as credible geological studies. One of the
most cogent papers contradicting Hapgood's theory is by Steve
Krause, who is apparently a German instructor at the University
of Wisconsin.[25] His paper has been copied and reprinted all over
the internet. Unfortunately he doesn't even mention the climate
data or the geomagnetic evidence from volcanic rock samples. His
argument against Hapgood's theory is mostly theoretical, explaining

reasons why it's unlikely to be true. I later found out that it appears Krause was an undergraduate German major at the time he wrote this paper. And there is a lot of similar unfounded skeptical views to be found on the internet.

Even Wikipedia, which of course is not to be confused with Encyclopedia Britannica, has an article that is extremely skeptical of Hapgood and earth crust displacement theory. But a lot of the most damning conclusions about his theory are lacking references. In fact Wikipedia has wisely added footnotes that say "citation needed." Even the external links at the bottom of the page are all skeptics' articles and sites, created to debunk Hapgood's theory. One of them, ironically, is a Creationism website, a theory significantly less scientific than Hapgood's.

Of course there are other articles and websites that are just as biased in favor of Hapgood's theory, but they are no more scientific or credible than the skeptics. Dan Eden, writing on Viewzone.com gives a supportive analysis of Hapgood's work.[26] Eden seems credible and fairly balanced, but Viewzone.com seems to specialize in paranormal phenomena and unsolved mysteries, such as the ever-popular image of Jesus appearing on toast. And Eden mainly just cites research from Hapgood's book. Followers of Graham Hancock are of course enthusiastic about earth crust displacement theory. One such researcher, a retired Italian Navy Admiral, Flavio Barbiero, is convinced that Hapgood's theory is correct and he focused on the theoretical cause of such a displacement.[27] But what about scientific evidence that it actually did or did not happen? Where is that research?

Probably, one of the best summaries of research on the Hapgood-Campbell theory of earth crust displacement is written

by F. N. Earll, of the Geology Department at Montana College
of Mineral Science and Technology. He wrote the forward to
the second edition of Hapgood's book (the first edition forward
was written by Albert Einstein). He wrote that there was a
predictably skeptical reaction to this revolutionary new theory,
but "their reaction could hardly be described as rational; hysterical
would be a better description."[28] He goes on to write that
"one observed, indignantly, that Hapgood was not a geologist.
Admittedly this is a cardinal sin but hardly one punishable by
scientific excommunication." He notes that "another cited, but
failed to name, a scientist whose findings conflict with those
of several world-renowned authorities selected by Hapgood as
sources of technical data, and used this lack of agreement as an
incontrovertible condemnation of the entire book."[29] Finally, he
states unequivocally that "nowhere in all that has been written
about the book, have I found a single authority who has calmly
and rationally offered a clear and documented criticism of the basic
theory involved. ...Frankly I wish someone would."[30]

Need for Further Research

Earll is obviously a supporter of Hapgood's theory. As a geologist
he states his professional opinion that, though he was skeptical
after the first edition of his book came out, "I personally feel that
in light of the data presented by Hapgood in this, the second
edition of his book, such (earth crust) dislocations are no longer a
matter of question."[31] He then complains of the lack of scientific
research either supporting or refuting Hapgood's theory or the
evidence presented in his book. He concludes his paper with the
following call for further research. "Now I ask – no, I implore – my

colleagues, those most competent to assume the task, to attack this theory with the weapons of well-documented proof."[32] He cites the dearth of data that I had also discovered and he pleads with his fellow research geologists, "let us not bury this idea prematurely through prejudice, as so many valuable ideas of the past have been buried, only to be sheepishly exhumed in later years." Earll finally and simply says of Hapgood's theory that, "if it is an unworthy thing let it be properly destroyed; if not, let it receive the nourishment that it deserves."[33]

Ironically Earll's plea echoes a very similar plea by Kirtley F. Mather, Professor Emeritus of Geology, at Harvard University. Like many others, he too was disturbed by the proposed theoretical cause of such a crustal displacement, and was unmoved by Einstein's support of it. He did not think unevenly deposited ice caps would produce the necessary force to move the crust. But he too thought that it was such a compelling theory that it deserved attention and a good deal of rigorous research, either to confirm or reject it. He wrote that, "the marshaling of data from many diverse fields of study and their interpretation in causal terms are sufficiently novel to make the author's ideas worthy of careful study and appraisal."[34] Mather concludes very much as Earll did that "the numerous unsolved problems to which Mr. Hapgood directs attention should be the subjects of intensified debate among scientists in every part of the world."[35] That was in 1959. Eleven years prior to Earll's impassioned plea for more research. It too was apparently unanswered.

Mather had urgently called for more research on this topic 53 years ago. And, as far as I can tell, his call for further research has been unanswered except for a few research papers here and there, such as those mentioned above. It appears that Earll's dire warning of prematurely burying the theory because of prejudice and then having to sheepishly exhume it years later, is exactly what has come to pass. Perhaps now, a serious examination of Hapgood's theory and evidence will be revived in the field of research geology and geophysics.

Conclusion

I have to admit that for a supposedly controversial theory, and one that has been mostly rejected by geologists, there is very little research to support such opinions. In fact, there is very little research at all, either supporting or rejecting it. Of course, I have quoted several highly respected geologists who support the theory, and at least one who at least obliquely challenges the findings of the magnetic rock samples. Overall, the weakest part of the theory has always been the lack of a credible cause of such an event. But with some mixed support for the jovian mass theory, even that is back on the table. Then there is a plethora of non-scientists who have weighed in, either pro or con. Of course, that's not scientific evidence, but rather an indication of his popularity or lack of it.

In the end I have no choice but to conclude that the earth crust displacement theory is a fundamentally sound theory, which is lacking the additional research evidence that would usually follow such an important theoretical breakthrough. For now, I think that there is simply not enough evidence to reject the theory, and more than enough evidence to support its plausibility.

12

EVIDENCE OF A JOVIAN MASS

In the last chapter I complained of a lack of scientific research on Charles Hapgood's theory of crustal displacement, outside of the extensive evidence that he presents himself. But even that would be a boon to this research-anemic theory. Even the main proponent, Cruttenden, uses mostly circumstantial and theoretical arguments to support his theory.[1] But there is some meager evidence to support this theory, and it comes from very credible sources.

Jovian Mass Research

As you will remember from Chapter 7, a jovian mass is any object with a mass as large as or larger than the planet Jupiter. We are primarily interested in the mass of the object not its size, so it could be very large and gaseous, or it could be smaller and very dense. This unknown object could be a planet or a very small, inert star. This object is almost certainly unknown to us now, though there is a theory that it may be a known star. And, though we don't know for sure that such an object exists, there is a good deal of mathematical and circumstantial evidence that such an object should theoretically exist.[2] Such an object is the only way we can explain a number of anomalous facts, such as the precession (wobble) of the entire solar system, the fact that precession is speeding up in accordance with Kepler's law of planetary motion, and the Sun's apparent missing angular momentum. And of course there is the Hapgood's theory, which explains ice ages, polar shifts, and much more, but that theory requires a force acting on the earth that is so great that it could only be produced by a jovian mass with a long elliptical orbit, passing through our solar system, near earth. So it makes sense that such an object might exist, but what proof is there that any such object really does exist? Have we discovered any such object? Well, maybe we have.

What's Flinging Comets?

In February of 2011, an article appeared in the journal Icarus, which provides "persistent evidence of a jovian mass solar companion in the Oort cloud."[3] That is in fact the name of the article. The researchers, John Matese and Daniel P. Whitmire, of the department of physics at the University of Louisiana at Lafayette, have put together compelling mathematical calculations based on a strange phenomenon that Matese had been noticing since 1999.

The strange phenomenon is the flinging of comets into our solar system from the outer Oort cloud. You will remember that the Oort cloud is the area of space that surrounds our solar system, and it is comprised of mostly rocks, ice, and frozen rubble, and it also contains most of our comets. The comets are on long elliptical orbits that takes them way out into the Oort cloud and then gradually returns them for dramatic fly-bys of our solar system, creating dazzling displays with their long bright tails. Well, according to Matese and Whitmire, something out there is flinging comets into our solar system.

It was always thought that comets were on regular orbits around the Sun, and that only occasionally are their orbits disrupted by collisions or near passes with other large objects such as asteroids. These disruptions were assumed to be random events, owing to a large number of small objects in the Oort cloud, which might be able to potentially disrupt their orbits. But Matese and Whitmire noticed something unusual, it appears that many of these comets are coming from the same place in the Oort cloud. This suggests that it may not be many small asteroids but rather one large mass, which is diverting many of these comets, from the same location in the sky.

The researchers present an impressive set of statistical equations in which they analyze all the new comets, listed in the 17th Catalogue of comets. They looked at the origin of the comets, their trajectory, and plotted them on charts. Then they performed statistical calculations to determine the likelihood of so many comets with similar orbits, originating from the same place. In a thorough array of bar charts, scatter-plot distribution charts, and linear charts, they graphically show the details of their analysis. Many of these charts clearly show the majority of comets conforming to a mathematical trajectory of where the comets should be coming from if there was a jovian mass in the Oort cloud flinging them this direction.

One chart clearly shows the confined pattern of the distribution of all new comets, correlating to the researchers thesis (this is a scatter distribution for aphelia directions of 17th Catalogue new comets, binding energy parameter of $x = 106$ AU/A, in the interval $30 < x < 60$ and tidal characteristic $S = -1$).[4] In a further statistical analysis of the confidence regions for orientation angles (Ω = galactic longitude of ascending node, i = galactic inclination), they clearly show that there is a large circle in the Oort cloud which contains a jovian mass with a 99% confidence level, a somewhat smaller circle with a 95% confidence level, and amazingly they find a very, very small point in the sky that they can say contains a jovian mass with 68% confidence.[5] This last finding may not seem like such a high level of confidence, but to be able to pinpoint the origin of so many comets to such a confined area of space, with such a high degree of confidence is pretty impressive.

There are many more statistical analyses, charts and diagrams presented. I would refer any curious scientists to check out the article on the SciVerse website, or in the journal Icarus in which it was published.[6] But for the average reader, the only way most people can evaluate these findings is to look for confirmation from other scientists and journals.

In May of 2011, Jack J. Lissauer of the NASA Ames Research Center joined Matese and Whitmire in presenting a paper to the 218th meeting of the American Astronomical Society (AAS), summarizing their data, and proposing that the detached Kuiper Belt object Sedna was likely flung by such a jovian mass.[7] They further conclude that this jovian mass is not just an orbital body, but also a binary companion to the Sun. So what did the scientific community say?

The Response to the Jovian Mass Evidence

The reaction of the scientists present at the AAS meeting was apparently mixed. Many were very impressed with the data, but others felt that more evidence is required. Hal Levison at the Southwest Research Institute in Boulder, CO, who has also previously authored a paper on the Oort cloud, stated: "Incredible claims require incredible proof," and he goes on to suggest that the statistical equations used should be examined and re-examined.[8]

Much of the controversy over these findings does not really lie so much in the statistics, but rather in their connection with the mythical planet Nibiru, or planet Nemesis as it is sometimes called. David Morris, Astrobiology Senior Scientist at NASA, echoes this sentiment. In the same article above, he was quoted as saying that even if this binary companion does exist, it is not Nibiru! He goes on to say that, "Nibiru has an orbit of 3600 years," and is going to rendezvous with the Sun in 2013. So they "could hardly be more different."[9]

Aside from the question of how this NASA scientist has so much hard data about an otherwise non-existent planet, his comments clearly do not refute the new research findings that were presented by Lissauer et al. Rather, he appears to concede that such an object may well exist, but it's just not Nibiru. Overall most scientists are rightfully hesitant to say that an unknown planet, much less the Sun's binary companion, really does exist until further proof is shown.

The good news is that the proof may soon be coming. Even as I am writing this, NASA is analyzing data from WISE, the Wide-field Infrared Survey Explorer. Matese and Whitmire say this NASA project could very well confirm or disconfirm the existence of a jovian mass when all the data is analyzed.[10] The problem is that it could take years to analyze all the data collected by WISE. Until then astronomers, physicists, and those curious about the ancient prophecies are anxiously awaiting the results.

Finally, you know you've made a ripple in the world of science when you make Scientific American. In May of 2011, this

esteemed science magazine ran a blog article and a multimedia pod-
cast, by science editor John Matson, on the Lissauer, Matese, and
Whitmire paper, which had been presented at the AAS meeting
that same week.[11] He gave a fair and balanced review of their work,
and appears as eager as anyone to see the results of the WISE study.
Matson refers to the jovian mass with its nickname Tyche, after
the Greek goddess of fortune and prosperity. The name was chosen
specifically so as to not be confused with Nibiru, or Nemesis. But
as the bard said, "a rose by any other name …" smells the same. As
the above quotes from scientists suggest, the majority of scientists
still think of Tyche as Nibiru, and for that reason alone they are
skeptical.

Let's Get Sirius

Walter Cruttenden, a leading proponent of the jovian mass/binary
theory, believes that Matese and others are looking in the wrong
place. He suggests that the Sun's binary companion is actually a
large, bright star. In fact, it is the brightest star in the sky, according
to Cruttenden. He believes that our Sun's binary companion is
Sirius, the Dog Star.[12] He bases his conclusions on the mythology
about Sirius, and on other mythological references to Nibiru.
He points out how the Orion constellation is important to so
many ancient cultures, and he points to Sirius's role as Orion's
mythological faithful companion. But he also bases his conclusions
on some scientific data as well.

He does list some interesting items. Specifically he says
that Nibiru means, "crossing," or "crossing star." And indeed Sirius
does seem to be moving and, if it continues on its path, it will

cross the entire sky, on a path that might well take 12,000 years. He calculates the distance from us, the mass, and other relevant features of Sirius, and concludes that this star fits the profile of the Sun's missing companion star. He suggests that our precessional movement, combined with the apparent movement of Sirius, will result in a rendezvous of the two star systems in about 10,000 years. But, this is nothing new; this observation was first made by mathematician, Egyptologist, and philosopher Rene Schwaller de Lubicz, over 50 years ago.[13]

Unfortunately there is only circumstantial proof that Sirius is our potential binary companion. I have seen no serious studies that produce actual evidence of such a probability. Other than the kind of calculations performed by Cruttenden and Schwaller de Lubicz, how would you go about proving something like that? The Sirius theory might well be right but we cannot neither prove nor disprove it at this point.

New NASA Data Supporting Binary Theory

There is one more piece of evidence. But, it's a bit circumstantial. In 2004 NASA launched a probe to study the effects of gravity. It was called Gravity Probe B. The satellite used four precision-engineered gyroscopes to measure the gravitational geodetic effect and the frame-dragging effect, both predicted by Einstein's theory of relativity.[14] Alas, the data came back as just "noise." It made no sense. They believed that this was due to solar flares in 2005. But the problem persisted. They concluded that the gyroscopes must be off. But, given that the gyros, some of the most accurate ones ever constructed, were specifically designed for this mission and checked by hundreds of scientists, and that it was a fairly simple experiment, it seems unlikely that the data would be that incorrect.

When the data was analyzed, it was found that there is only one way that this data could possibly be right, and that is if the solar system was curving through space due to the gravitationally pull of another nearby star. Specifically, if our sun were part of a binary system, then the data would make perfect sense.

The managers of the Gravity Probe B asked for additional funding to try to extract more information, but a review panel doubted they could succeed in their mission, given the "large" discrepancies in data coming back from the probe. The review panel concluded that the data coming back from the probe would have to "overcome considerable scepticism in the scientific community", and they canceled the program.[15]

No one from NASA ever came out and said that the data was evidence that our sun was part of a binary system, and the primary scientist on the project, Stanford University physicist Francis Everitt, never said as much either. To this day, they say that the data was just "noise." But, any physicist looking at the data would have to concede that the results would be consistent with a solar binary system. Ironically, the data could probably have told us exactly where to look for a jovian mass, if the program was allowed to continue.

Summary of Evidence

We already knew that scientists could not rule out the presence of a jovian mass without mapping every cubic mile of the vast Oort cloud. But, just in the last few years there is actually some very supportive evidence that there really is a jovian mass existing out in the Oort cloud, nicknamed Tyche. Furthermore, the evidence suggests that this object may be a binary companion to our Sun. This evidence does not come from quacks. Serious researchers at NASA Ames Research Center and the University of Louisiana have presented these findings at a major scientific conference, and this evidence was also published in a respected journal.

The scientific community, while cautious, is eagerly awaiting possible confirmation from the NASA WISE sky survey. Meanwhile, there is anomolous data from a NASA gravity probe that is consistent with our Sun having a binary companion.

All in all, it seems that the biggest objection scientists have to the possible existence of a jovian mass is the fear that they might have proven that the mythical planet Nibiru really does exist. And this seems to be a fate worse than death to many scientists.

When you take into account all of the above factors, I have no choice but to conclude that there is serious, credible evidence supporting the idea that our Sun has a binary companion, which is a jovian mass far beyond our solar system, with a long elliptical orbit. And if this object entered our solar system, then it could very well cause a disruption to the earth that is consistent with the worst accounts of cataclysms found in ancient mythology. Further, it would only make sense that such an object would enter our solar system as part of its orbital path, seeing that this is exactly the same pattern that we see with other objects from the Oort cloud, namely comets.

13

EVIDENCE FOR REISER'S THEORY

Finally, we come to Oliver Reiser's evolutionary theory. Remember from Chapter 8 that, while writing from the 1930's through the 1950's, Reiser had many fantastic ideas that he deduced from scientific information available at that time. And many of his apparently crazy theories and predictions have actually come true. Regarding his evolutionary theory, he maintained that we do not evolve gradually over time, but rather in periodic events which coincide with the cycle of the precession of the equinoxes. He felt sure that evolution was propelled through DNA resonance vis á vis the cosmic radiation that is allowed to strike earth due to fluctuations in the earth's geomagnetic field. This would probably be caused by movement or a complete reversal of the magnetic poles. Thus genetic mutations would occur as a result of these periodic geomagnetic polar fluctuations.

Reiser seemed to have a teleological interpretation of evolution, bordering on what we would call "intelligent design." He had a deep reverence for spirituality especially in eastern philosophy. He believed that we as a species were progressing toward a synthesis of western science and eastern mysticism. The result, he believed, would create a hyper intelligence with a higher consciousness and a world unity of humanity. He even believed that we were moving toward a global consciousness, perhaps even so far as saying one global mind, made up of billions of cells, each cell an individual person.[1] But what proof do we have?

The proof of this is the dodgiest of them all, yet not non-existent. In fact it was evolutionary and anthropological evidence, which lead him to the above deductions in the first place. You see, Darwinian evolution maintains that every once in a great while a mutation occurs in a species. If that mutated species is better adapted to the environment, then it will be more successful than others. In this way the species will evolve. So, Reiser simply asked the obvious questions. Why don't we see more of these mutations? And what would cause a genetic mutation in the first place? Informed by the research done on survivors of Hiroshima and Nagasaki, he concluded that radiation could cause mutations, and of course our solar system is filled with radiation. But what would cause that radiation to break through our magnetosphere to affect our DNA? He concluded that there must be fluctuations in the geomagnetic field, which then causes fluctuations in the amount of cosmic rays striking the earth. By attempting to answer these questions, he arrived at the above theory. It was derived from Darwin's theory, and from the anthropological and paleontological evidence of intermittent bursts of evolutionary progress.

Most scientists acknowledge that cosmic rays such as UV radiation and X-Rays can affect DNA. In fact, the famous astronomer Carl Sagan claimed that human evolution was likely the result of incoming cosmic rays, possibly from some distant neutron star. [2] Researchers such as science writer Andrew Collins has gone on to suggest that bursts of these cosmic rays can be detected in ice core samples in Antarctica, and they correlate with advances in human evolution.[3] Though he too believes that these bursts are probably from some distant neutron star, such as Cygnus X-3, aiming a beam of radiation directly at Earth. So Reiser's theory, at least indirectly, has some support in the scientific community.

The problem with Reiser's theory is that there is a strong theoretical objection to this theory, and that is that our atmosphere and ozone layer block out most UV and X-Ray radiation. They say that the ozone is a more relevant factor than the shield-effect created by the earth's geomagnetic field.[4] This shield is believed to keep out only charged particles, driven by the solar wind. But it is argued that our atmosphere would stop them anyway.

The problem with this objection is that it is primarily theoretical, and there are theoretical counter arguments to be made. Collins acknowledges that cosmic rays are deflected by the atmosphere, and goes on to point out that when cosmic rays hit the upper atmosphere they break up when they collide with the nuclei of oxygen and nitrogen, creating charged secondary particles. Some create isotopes that are preserved in lakebed sediments and in ice, which is how we can measure the amount of cosmic rays striking the earth in the past. [5] He specifically cites Beryllium-10 as an isotope found in ice core samples, which indicates the presence of increased cosmic radiation striking the earth in the past. So, a gap in the protection of the geomagnetic field, in theory at least, would increase the amount of charged secondary particles striking the earth.

These are still theoretical arguments. It would be important for scientists to witness a magnetic pole reversal first-hand, to study it and collect data, in order to determine the full effect of such an event. The truth is that we may have some data to analyze, but it doesn't come from astrophysics.

The Upper Paleolithic Revolution

Fortunately, to find the proof of this kind of punctuated evolution, as Reiser's theory suggests, we need only look at human development over the last 60,000 years. Was there in fact a leap of evolution found in the artifacts of early people? The answer is yes. Something did occur thousands of years ago. Many anthropologists would say that there was not one, however, but two different leaps in human evolution. These are referred to as the Upper Paleolithic Revolution[6] and the Neolithic Revolution.[7] The first of these, the Upper Paleolithic Revolution, is the time when our human ancestors started acting more of what we might call human. It is widely acknowledged that 45,000 – 50,000 years ago we see the first art, idolatry, formal burials, musical instruments, and many other features of modern human culture. Based on the coordination of their activities, it is likely that they had some effective form of language as well. This was the time when our primitive ancestors literally went from being grunting cavemen to being early humans.

It should be noted here that when we say evolution we are not talking about major differences in our physical appearance, nor even our brain size. We are really talking about the sudden appearance of human cultural objects, not previously seen in the archaeological evidence. Cavemen made decent tools for a hundred thousand years or more, with no attempt at art. Then suddenly, about 45,000 to 50,000 years ago, all over the world people started making art, music, and idols. You might think that this would be a more gradual development that began around 45,000 years ago and progressed from there. But the data from artifacts do not indicate such a pattern. The art and artifacts from this period are amazingly consistent across cultures and locations, and remained so for tens of thousands of years. We do not see a learning curve but rather a plateau of development. Then this all changed about 12,000 years ago.

It is very interesting that anthropologists would use these dates, to mark these changes in human development. Think of what else might have happened 50,000 years ago, and again 12,000 years ago. What's so special about these dates? These are the exact dates that Hapgood believes that two major earth crust displacements occurred.

Neolithic Revolution & Göbekli Tepe

Though the Upper Paleolithic was marked by the creation of art and idolatry, and goddess figures are found all over the world from this period, still there was no buildings, no monuments, no written language, no intricate tools, or advanced technology of any kind. Then, again we see another jump. Around 12,000 years ago, we have the earliest evidence that people began creating civilization, with buildings, temples, agriculture, pottery, metallurgy, assorted technology, and astronomy. Archaeologists and anthropologists call this the Neolithic Revolution.

How rapid was this progress? A good example is found at Göbekli Tepe. According to an 2008 article in *Archaeology*, recent archaeological digs at this site in southern Turkey, near the fertile crescent, might very well reveal the earliest buildings in the world.[8] Previously, with the finds at the Tower of Jericho, it was thought that the Neolithic Revolution occurred around 10,000 years ago, but with this find and the even more recent finds at Wadi Faynan in Jordan, which date to 11,700 years ago, archaeologists have had to push back that date considerably.[9]

Dating of charcoal found at Göbekli Tepe reveals that the site was occupied as much as 12,000 years ago. But they were not small or common structures. According to Klaus Schmidt of the German Archaeological Institute of Istanbul, the site appears to be a temple complex, covering 90,000 square meters, with 17 chambers, and elaborately carved stone pillars up to 18 feet in height, weighing up to 16 tons each.[10] The buildings feature terrazzo flooring, numerous stone carvings of animals and faceless humanoid forms, and rock masonry walls. But this was not the end result of thousands of years of architectural development. This amazing structure, apparently a temple of some sort, was the earliest building we can find. This is the oldest temple in the world. In fact, the later structures at this site are smaller and seem less sophisticated, the reverse of what you might expect.

These are not the only early Neolithic monumental structures. Robert M. Schoch, a geologist at Boston University, has suggested that Great Sphinx of Giza must be much older than earlier estimated. He has based his findings on the weathering of the stone carving. He believes that the Sphinx is at least 7,000 years old and may be quite a bit older.[11] Other researchers have commented on the fact that the bottom of the pyramids contain very different stones than the rest of the pyramid. They are much larger, and are of a different quality. They suggest that the pyramids might have been rebuilt during the reign of Khufu and Khafre, but may be thousands of years older than previously thought. Writer and researcher Robert Bauval, using sophisticated computer software, has discovered that the alignment of the stars with the Great Pyramids and the Sphinx line up perfectly with the night sky of about 12,000 years ago.[12] And, as mentioned earlier, Professor Arthur Posnansky of the University of La Paz, like his colleague Professor Rolf Muller, is convinced that the ancient city of Tiahuanaco was first inhabited as long as 12,000 years ago.[13] So there is actually ample evidence that Neolithic structures may have been springing up all over, and all at about the same time.

Since about 12,000 years ago, no one disputes that our ancestors were fully modern humans, with all the same intellect, talents and capabilities of people today. And again it consisted of a plateau of development. At this point humanity appears to have arrived at our present state of mental and emotional development. From then on, our cultures and civilizations progressed not as the result of genetic evolution, but as mimetic evolution.

Mimetics vs. Genetics

Mimetic evolution is the evolution of learned behavior. Language, science, technology, literature, cultural norms, history, and folklore are all part of mimetic evolution. And just as the building block of genetic evolution is the gene, the building block of mimetic evolution is a mime, that is something that is learned or copied from someone else. It could be as simple as learning to build a fire, or as complex as the invention of the number zero. Eventually it snowballed and resulted in modern civilization. So arguably the last 12,000 years of evolution has all been mimetic.

Stepwise Progression

The million dollar question is what created this stepwise progression in human evolution? For millions of years we evolved with no modern human culture at all, that is no art or religion, or science. Suddenly we start making human art and artifacts. And we do this for tens of thousands of years, with no change or progress in the quality of artifacts for all those years. Then, out of nowhere, 12,000 years ago we suddenly start building huge, stone temples, settlements, creating cultures and then civilizations. When you look at it like this, the Upper Paleolithic Revolution was really an interim plateau, punctuated by two leaps in progress, one at the beginning and one at the end. Somewhere between 45,000 to 50,000 years ago, we start to see signs of art and ritual, of beliefs and culture. Based on artifacts, we can say with confidence that it appears that this level of human development continued without change until about 12,000 years ago. Then, there was another jump. From that point on we start to see the artifacts of modern human prehistory, and the roots of the current peoples of the world.

There is another interesting set of data that fits with all the above research like a missing jigsaw puzzle-piece. Actually it's not really a set of data so much as a lack of data. You see, after this apparent leap in human evolution at the beginning of the Neolithic Age, we find a large gap in monumental architecture or civilizations of any kind. There are a few examples such as Old Europe, and the Varna and Çatalhöyük cultures. But there is largely a huge gap in artifacts of advanced civilizations, until about 5,100 years ago. Then, as mentioned before, civilization seemed to bust out everywhere at once, right about the time that the sea levels rose hundreds of feet, all over the world.

In retrospect it now seems obvious that the artifacts documenting this leap in evolution at the end of the Pleistocene may have been partially masked by a tendency for our Neolithic ancestors to build their civilizations in low-lying, coastal areas. These were probably lush plains, capable of producing high quantities of food, and providing easy access to the sea for efficient transportation to other regions and civilizations. It now appears from the research of Jeffery Rose and others that 7,000 years of progress was probably erased by cataclysmic flooding and seismic instability that occurred between 5,000 and 6,000 years ago. This created a pattern of amazingly advanced civilizations springing up all over the earth at the same time, on higher ground but near the flooded plains.

Conclusion

Of course there is much, much more evidence of genetic mutations occurring out of nowhere to be found in paleontology textbooks. It is precisely these curious jumps in evolution that fuel the arguments of Creationism and Intelligent Design, which point to the lack of missing links and prevalence of these jumps in evolution.

We focused primarily on human evolution because that is the topic mentioned in the ancient myths and prophecies. They say that with each age of humanity we grow and develop. And they say that 2012 will mark a leap forward in our spiritual and intellectual development. From the sound of it they are describing genetic, not just mimetic evolution. The difference is that mimetic evolution gradually, albeit exponentially, builds while genetic evolution tends to create sudden leaps of progress.

In conclusion, we studied a scientific theory by Oliver Reiser, which would account for these sudden leaps in genetic evolution. We then showed that there is a clear and distinct pattern of artifacts from the last 50,000 years, showing two distinct leaps in progress. They are so well known in the field of anthropology that they are named the Upper Paleolithic Revolution and the Neolithic

Revolution. And when we analyze this pattern we see that it reveals characteristics consistent with Reiser's theory.

Most astonishing is the fact that this evidence for Reiser's theory is completely compatible with the timing provided by Hapgood's theory. He asserts that earth crust displacements occurred 50,000 years ago and 12,000 years ago.[14] Reiser asserts that any fluctuation or movement of our geomagnetic poles, especially polar reversals, would produce genetic mutations, and thus evolutionary leaps. And at these precise points in prehistory, that is exactly what we find in the archaeological evidence.[15]

We know a cataclysm occurred at the end of the Pleistocene, approximately 12,000 years ago.[16] Hapgood asserts that we had our last earth crust displacement about 12,000 ago.[17] Now we know that anthropologists believe that the Upper Paleolithic Revolution ended and the Neolithic revolution began about 12,000 years ago, and that is when modern humans began roaming the earth.[18] This seems to be more than just coincidence. Again, the pieces of evidence just seem to fall together like a puzzle, each piece perfectly abutting and joining with the other pieces.

Given all this evidence, and little or no contradictory evidence, we would have to conclude that Reiser's was probably on to something. In short, even if he did not have the DNA mutation mechanism completely correct,[19] his theory perfectly fits the data that we have about human evolution in the last 50,000 years, and coincidentally it just happens to also fit with Hapgood's crustal displacement theory. The time periods are not just close, but dead on. In fact, it actually relies on Hapgood's theory, because without evidence of movement or disruption of the poles, especially the magnetic poles, Reiser's theory would fall apart. Once again, the evidence indicates this jigsaw puzzle effect, where each bit of evidence perfectly dovetails with other bits of information, and with no discrepancies. Ironically, although both Reiser and Hapgood collaborated with Albert Einstein, the two men's research was completely independent of each other, making the above coincidences all the more incredible.

We have to wonder if Einstein, having intimate knowledge of both Hapgood's and Reiser's research, was able to put these pieces together before his death. Who knows, if he had knowledge of the ancient Mayan prophecies, he may well have become the greatest disaster theorist of all.

Part IV

SUMMARY & CONCLUSIONS

14

APOCALYPSE NOW?

Let's sum up all the information presented so far. We have now looked at Mayan mythology, their history of previous cycles of creation and destruction, and their prophecies of doom. We found strikingly similar mythological accounts and prophecies from around the world, and going back thousands of years. A summary of all the myths and prophecies from around the world is roughly as follows.

Summary of Mythology and Prophecies

- There have been many ages of humanity, and each time we build civilization it is destroyed by natural disaster.

- One of the greatest of these cataclysms involves a complete change in the path of the Sun in the sky, resulting in it dawning in a different direction.

- This was a cataclysmic event that involved both fire and flood, and a burning, black rain, and a long period of darkness.

- The last of these cataclysms primarily consisted of a great flood that affected the entire earth, wiping out civilizations, and apparently "sinking" cities into the sea.

- It is also mentioned in myths that this event is because of the movement of heavenly bodies.

- [ ] Many prophecies say that we will know when the next cataclysm will strike when we find a new star in the sky.

- [ ] It is foretold that this will happen again someday.

- [ ] This cataclysmic event is also said to be the start of a new age of humanity, one in which we will grow and advance spiritually, to become more than what we are, and with greater world unity.

Summary of Scientific Theories

The abundance of scientific theories is immense and confusing, but out of the chaos there have emerged several theories with some merit. We only looked for theories which could account for the specific cataclysmic features found in the above mythology. And we limited it further to only those myths that were corroborated by other myths from other cultures. When we were finished, there were only 3 theories that made the cut. Of course we looked at Hapgood's earth crust displacement theory, Cruttenden's jovian mass/binary theory, and Reiser's geomagnetic/evolutionary theory. What follows is a summary of these theories.

Earth Crust Displacement Theory

- [ ] Charles Hapgood's earth crust displacement theory explains the cause and history of the ice ages, and also previous cataclysms.

- [ ] He did not know it, but his theory also explains the many myths, from different cultures, of the Sun changing its path in the sky.

☐   The only phenomenon that could cause all these specific changes at once would be if the crust of the earth slipped, or was displaced, relative to the mantle underneath.

☐   In this scenario, the earth crust would slip such that the north pole might move as much as 30° latitude, producing cataclysms, flooding, major climate changes and possibly many extinctions.

Jovian Mass Theory

☐   Walter Cruttenden and others have proposed a jovian mass / binary companion theory. He suggests that we are in a binary system, and our Sun orbits another large object, such as a small star or a jovian mass. This is based in part on anomalies in the observations of precession mentioned below.

☐   One of the latest discoveries in the study of the precession of the equinoxes, is that it may not be specifically an earth wobble, but rather a solar system-wide wobble.

☐   New observations indicate that recently precession has begun accelerating, and this acceleration conforms to Kepler's law of planetary motion.

☐   Calculations also show discrepancies in the angular momentum of the Sun, which can only be fully reconciled if the Sun is in a binary system.

☐   The length of one orbit with our binary companion would be equal to 1 precessional cycle, or about 24,000 years.

☐   And, when the object comes closest to the Sun, great changes may occur on earth.

Reiser's Evolutionary Theory

- ☐ Oliver Reiser developed a theory, which explains sudden bursts in human evolution as being due to fluctuations in the earth's magnetic field and movement of the magnetic poles.
- ☐ Based on his study of evolution, he believed that there is a teleological direction to our evolutionary progress, and that the next advance will likely involve psychic or spiritual growth, higher consciousness, and world unity.

Synthesis of Theories

Amazingly all of these scientific theories fit together to create a synthesis, a composite theory comprised of several different theories. For instance, Hapgood's theory requires some great force acting on the earth's crust to cause displacement. For years the lack of any such force was one of the biggest objections to his theory. Most scientists agree that only a massive object larger than Jupiter, a jovian mass, passing near earth could cause such an event. So Hapgood's theory needed the jovian mass theory to complete it.

Now, new information about the precession of the equinoxes suggests that the phenomenon may be the side-effect of our Sun being in a binary orbit with a jovian mass, thereby validating the jovian mass theory. Finally, Reiser's theory of periodic advances in human evolution requires some movement or a complete reversal in the poles to create this effect. Fortunately, Hapgood's research found geomagnetic evidence of multiple polar displacements, and at least two complete magnetic pole reversals, just as Reiser had predicted. These theories seem fit together like pieces of a jigsaw puzzle, each one filling in discrepancies in the other.

Summary of the Evidence

Lastly, we looked for any evidence, which specifically supported or contradicted either the mythological accounts of the past or the scientific theories of future events. We looked at the geological record, the climate record, the paleontological record, the archaeological and anthropological records. Then we looked at the latest astronomical research. I think it's important to note that we did not 'cherry-pick' data that supported a doomsday scenario. Rather we tried very hard to find actual evidence contradicting the above myths and theories. A summary of the evidence is as follows.

<u>Evidence of Mythical Accounts</u>

- ☐ In examining all of the evidence, we see that sea levels did indeed rise all over the world, just prior to the beginning of recorded history, which theoretically would have created massive global flooding, just prior to 3,100 BC.

- ☐ There is currently growing evidence of multiple civilizations, which were actually destroyed by flood at that time, being overtaken by rising sea levels, or apparently "sinking" into the sea.

- ☐ There is even compelling evidence that both the myth of Atlantis and the myth of Gilgamesh were based upon real civilizations, which endured cataclysmic flooding and other catastrophes in prehistory.

☐ There is considerable evidence of a massive catastrophic event about 12,000 ago, at the end of the Pleistocene, in which the ice age suddenly ended, millions of animals died, and many species of animals went extinct. There is also evidence of major seismic upheavals at this time.

☐ Finally, there is evidence of human civilizations existing at that time, which appears to have survived and continued on into recent prehistory, thereby being capable of passing down eyewitness accounts of the last catastrophic event for thousands of years, through stories and myths.

Evidence of Hapgood's Theory

☐ Hapgood presented exhaustive evidence in his book, supporting his theory of earth crust displacement, this included climatological and geomagnetic evidence, as well as archaeological evidence.

☐ While many scientists have refuted his theory, they are largely theoretical arguments and most just dismiss the theory without conducting further research.

☐ It seems that his earth crust displacement theory may have been eclipsed by the popularity of plate tectonics theory, which came out at about the same time.

☐ There has been little or no research refuting the specific evidence that he presented, especially the climate evidence, which most scientists seem to concede, though they cannot explain it.

Evidence of a Jovian Mass

☐ The jovian mass theory has very recently, as of 2011, found support from NASA scientists and University physicists who claim to have found some unknown massive object outside of our solar system, possibly in an elliptical orbit with the Sun. Possibly a binary companion to the Sun.

[] This is in addition to theoretical evidence based on the relative angular momentum of various objects in the solar system, suggesting a binary system.

[] Careful observation of the effects of precession confirm that it is confined to objects outside of our solar system, suggesting it is caused by being part of a binary system.

[] There is evidence that confirms that the rate of precession is indeed increasing, consistent with Keplar's law of planetary motion, suggesting that our solar system is in orbit with a another object.

[] Finally, NASA probe B was shut down because the data was so unusual that they considered it was broken. But, the data would actually make perfect sense if our sun was part of a binary system.

Evidence for Reiser's Theory

[] Archaeological and anthropological evidence has revealed two relatively sudden leaps in human evolution, the Upper Paleolithic Revolution around 50,000 years ago, and the Neolithic Revolution, around 12,000 years ago.

[] Coincidentally, these dates are the same time periods when Hapgood estimated that there were previous polar shifts.

[] The above evidence, when combined, clearly supports Reiser's theory.

[] Most scientists are critical of the process by which Reiser theorized that this might have happened. Specifically they are skeptical that a geomagnetic reversal could increase cosmic radiation.

☐ We do, however, have evidence that UV radiation does affect DNA, and tentative evidence that electromagnetic fields do create altered states of consciousness.

How Many Coincidences is Beyond Coincidence?

First let's acknowledge that myths are just myths. They are neither true nor false, they are not scientific evidence, but just old stories. They may or may not be based on real events. Likewise, theories too are just theories. We don't know if they have merit until we look at the evidence. So, of all the above sections presented in this book, it is the evidence that is the most important. It is the evidence, which indicates if the myths and theories have any basis in fact or not. Without evidence, all of the above myths and theories are just so much hot air and spilled ink. Ok, so let's analyze the evidence.

I have to admit I was more than a little skeptical when I began. I am not pathologically skeptical. That is, I'm an open-minded skeptic, so I acknowledge credible evidence when I find it, no matter how bizarre the implications may be. But I really didn't think I'd find as much as I did. The thing that amazed me the most was the concordance of different theories, sources of evidence, myths from different cultures, all lining up in a truly uncanny fashion. The concordance of dates for the Upper Paleolithic Revolution and the evidence for earth crust displacement at that same time is a bit of a coincidence. Then the concordance of the Neolithic Revolution with the evidence of another crust displacement at the same time is a slightly bigger coincidence. But then when you look at the concordance of dates for the Neolithic revolution and the catastrophic end of the Pleistocene, combined with the geomagnetic and climate evidence for the last earth curst displacement, that's an even bigger coincidence. Then when you look at the likely effects of earth crust displacement and then look at mythological accounts of past cataclysms, the fit seems beyond coincidence. The odds of all of these coincidences just being due to random chance seems increasingly implausible.

Then let's consider the presence of heavenly bodies associated with catastrophes in ancient myths and prophecies; now we find that the only compelling theory that scientists say might conceivably cause an earth crust displacement is a jovian mass passing through our solar system. Then there's the coincidence that, oh yes, NASA scientists think they've just discovered evidence of a previously unknown jovian mass on an elliptical orbit with the Sun. At this point, you'd have to be pretty disingenuous to say that this is just a coincidence. And we have only skimmed the surface.

Within each of these general topics, such as displacement theory or jovian mass theory, there are a wealth of facts and details, dozens and dozens of which all coincidentally point in the same direction. This includes all the evidence of a cataclysmic end to the Pleistocene, all the varying climate evidence indicating a different location of the polar ice caps in the last ice age. Then there is all of the different myths from various cultures, from different continents, all with the same, very specific features. At a certain point, you'd have to admit that the evidence is far beyond what you'd expect from random chance, alone.

We have at least three well developed scientific theories that fit together like a jigsaw puzzle, tons of evidence to support each of them, and a concordance of global, ancient mythology that describes them. How many coincidences is that, 100? 200? More? One coincidence is a coincidence, two coincidences is also possible, three maybe. But after 23 coincidences in a row, you start to suspect that it may not be due to just coincidence alone. And, after about 101 coincidences, you'd have to be pathologically skeptical not to acknowledge that you just might have discovered something, even if you don't exactly know how it all works yet.

A Problem of Timing

As I said, I began this research by investigating the Mayan 2012 prophecies. And, the reason why I was able to predict that there would likely be no cataclysm in 2012 must be clear by now. While we may have evidence of global flooding just prior to the beginning of the last Long Count cycle (about 5000 years ago), there was just no evidence that *that* kind of event would be repeated in 2012. And, the 2012 date did not seem related to the other details in the prophecies, which describe a possible crustal displacement and a possible Jovian mass in orbit with our Sun. But there is evidence of a global cataclysm, which quite possibly could be able to produce all the different details we see in the mythology. The problem is it seems to have occurred 12,000 years ago! That does not seem to be related to either the Mayan Long Count or the 24,000 year long precession of the equinoxes.

The one thing missing in the evidence of past cataclysms is a meaningful timeline, which would implicate 2012 as a date of a recurring cataclysm. In other words. Even if we can prove that there was crustal displacement 12,000 ago, or thereabouts, what does that have to do with 2012? Two Long Count cycles ago was 10,254 years ago, three Long Count cycles ago was over 15,000 years ago. One precessional cycle ago was approximately 24,000 years ago. So none of these dates are in any way related to an event 12,000 years ago. In fact, the only item that was *not* supported by the evidence was the date of 2012.

The bad news is that the earth could be destroyed anytime, and by any number of other scenarios besides crustal displacement or a Jovian mass. Of course that doesn't explain how our ancestors might have predicted such an event, outside of astrology or reading tea leaves. Also it does look like there is some tentative proof that our doomsday scenario did happen about 12,000 years ago, so it probably could happen again in the future, possibly in another 12,000 years, if our doomsday synthesis of theories is correct. That's not nothing. If the world is destroyed again in 12,000 years, then we now have a pretty good idea of how that might happen.

The Second Blow

Our synthesis theory is that a jovian mass, which is a binary companion to our Sun, is the cause of earth crust displacement. So it would seem to me that in order for there to be a clear relationship between earth crust displacement and binary theory, they should both fit within a common cycle of time. That is if we can show that a polar displacement occurs every 24,000 years like clockwork, and we believe that our binary orbit is 24,000 years long, then you'd have strong support for our synthesis theory. Unfortunately, the evidence shows no connection between the two time-lines.

Hapgood collected evidence that showed crustal displacement occurs at relatively random intervals. Yes, a couple of the intervals are about 24,000 years apart, but the two most recent displacements that he found were somewhere around 38,000 years apart, +/- 2000 years (from about 50,000 years ago, to about 12,000 years ago). Even stranger, he presents climate data in his book that shows that the earth has had fairly consistent patterns of periodic ice ages throughout the Pleistocene Age, but that cycle looks to be about 36,000 years long. It is very hard to reconcile these dates. If we really grasp at straws, you could say that the distance between last two ice ages, and the length of the earth's climate cycle are both about 36,000 years. But, that really could just be a coincidence.

I do see one other feature that seems unusual. When you look at all the data presented by Hapgood, both geomagnetic and climate data, combined with all the other theories and myths, you see three relevant time periods, over and over again, 12,000, 24,000, and 36,000 years respectively. The last major catastrophic event was 12,000 years ago; the Vedic myths suggest that we are on 24,000 year cycle, which is comprised of a 12,000 ascending cycle and a 12,000 descending cycle. The precession of the equinoxes is actually 24,000 years long. Some of the cycles of earth crust displacement found by Hapgood reveal approximately a 24,000 year cycle. And finally, the second most common span of time between earth crust displacements appears to be about 36,000 years. Coincidentally, there is data that shows that the earth has had a very stable repeating climate pattern throughout the Pleistocene, on about a 36,000 year cycle. So? What does this mean? I don't know, but it does seem interesting that the same numbers keep popping up, and always in multiples of 12,000.

One could argue that the one bit of information that does not correspond to this apocalyptic numerology is the Maya Long

Count of 5127 solar years. But I do think it important to point out that 5 Long Counts equals 25,635 years. Why is that important? Because that's about how long the precession of the equinoxes was estimated to be 100 years ago, prior to the apparent acceleration of precession. So, it could be that the Maya accurately measured precession at the time that they were observing it, and just had no idea that it would begin to accelerate.

The Mystery of 12,000

I wish I could say that I have it all figured out. The numbers do not seem to be completely random. But I will leave it up to astrophysicists and catastrophe science geeks to ponder the mystery of the number 12,000. I do have some ideas. Suppose the idea of the Vedic yugas are right. Suppose there are two 12,000 year cycles, which combine to form a 24,000 year cycle. And suppose this process is not a nice, simple, repeating pattern. Let's say that in every other pass by the solar system the jovian mass is a little closer or a little further away from earth. It could be that the jovian mass (Tyche) swings by earth every 12,000 years, as part of a 24,000 year cycle.

To see how this works let's imagine that you are right in the middle of an elliptical NASCAR race track. It takes 2.4 minutes for a car to complete one lap. And we'll assume that there is only one car on the track. That means that every 1.2 minutes the car zooms past you, in the middle of the narrow infield. But let's say you are moving back and forth, from one side to the other. So sometimes the car zooms very close to you every 1.2 minutes if you time it right. Other times it zooms past you every 2.4 minutes if you stand still. Now let's say you had a sound recorder on you, recording sound not unlike the way we record climate patterns in the Pleistocene. Overall, you might hear a pattern of a loud, zooming race-car repeating every 3 half laps or every 3.6 minutes. If your movement in the infield is just right.

So, how could this happen with the earth, the Sun, and a binary companion to the sun? Well the location of the earth would likely be in a different place every time the jovian mass swung through the solar system. That's like the guy in the infield of a racetrack moving back and forth, from side to side, in a circular motion. Of course, the Sun would be at one end of the ellipse, so that means that the earth too would be at one end of the ellipse. To use our analogy above, that should put our bystander wandering in a circle at one end of the infield, but I think the timing would work better if we were in the middle of the elliptical orbit. With a person in the infield, you could theoretically get a cycle of 3.6 minutes, with nodes at 1.2 minutes and 2.4 minutes. That corresponds to a cycle of earth destruction of 12,000, 24,000, and 36,000 years, respectively, involving a jovian mass, and the variable position of the earth.

This might explain the apparent random time spans between earth crust displacements that Hapgood found. Of course the earth would be in a different location relative to this binary object each time it came around. The common denominator of 12,000 years indicates that there is a regular orbital pattern for the object, but the variability of crustal displacement cycles is due to our position relative to the object. Think of Halley's comet. When it came by in the early part of the 20th century it was a spectacular sight from earth. But when it returned in 1986 it was small and faint; what happened? It looked different because we were in a different position relative to the comet. The length of one orbit stayed the same, but the brightness of the comet might have a cycle of two or three orbital cycles. Now imagine that it's not a comet but the jovian mass, named Tyche, by NASA. We might very well have a completely different experience each time it comes around, depending on our location in relation its orbital path.

The Probability of a Future Cataclysm

Based on all of the above evidence, I think it unlikely that all these myths, theories, and data fit together like hundreds of coincidences, by random chance alone. It seems likely that we have, perhaps accidentally, discovered some natural phenomenon of which most people were completely ignorant. The most disturbing thing about my research is the apparent lack of interest, and the commensurate lack of rigorous research on this topic. You'd think that researchers would be getting huge grants, working their tails off, trying to see if there is anything to this potential doomsday scenario. Even if there is only a one in a thousand chance of a global cataclysm, that seems like enough of a risk to warrant serious research. We definitely need more research. We cannot keep dismissing this topic as being synonymous with the search for aliens and Bigfoot.

Fortuneately, based on all of the research presented here, I would have to say that a major cataclysm does not seem likely any time soon. In Hapgood's own research, two crustal displacements have never occurred a mere 12,000 years apart, it is usually 24-36,000 years apart. The climate patterns of earth are a consistent 36,000 years apart. So that's encouraging. But likewise we would be foolish not to take note of the fact that the last global cataclysm was exactly 12,000 years ago. That opens up a small possibility, if

Hapgood's theory is correct, and if Tyche is really lurking in the Oort and heading this way, and if the timing and placement of earth is exactly the wrong place at the wrong time. But that's a lot of ifs. If my NASCAR analogy is correct, then it is more likely that, at worst, we will have a non-cataclysmic glimpse of our Sun's binary companion, Tyche, sometime in our lifetime. After all, such an event would still be worthy of prophetic warnings by our ancestors, even if it proves to be less tragic than the last rendezvous with Tyche. And, such an event might well produce some strange and unforeseen effects in our culture and the course of civilization, fulfilling both Reiser's theory and many of the ancient prophecies.

The Final Word

I still think that future cataclysms, such as the one that occurred at the end of the Pleistocene or the one that wiped out the dinosaurs millions of years ago, appear completely inevitable in the future of the earth. And based on the theories and evidence presented in this book, I think we now have pretty good hypothesis of how one such cataclysm might occur. And, it looks like we might well have over 10,000 years to prepare for it. You may think that is a lot of time, but to protect billions of people from a global catastrophe of biblical proportions, that may be just enough time. We owe it to our descendants to investigate this topic rigorously, and to move toward a society that could one day face this epic challenge and survive.

Well that's it. Mission completed. Doomsday averted! I think we can all breath a sigh of relief now, and just enjoy the positive blessings that the ancients foretold for us, and our auspicious future.

15

WHAT TO EXPECT, WHEN YOU'RE EXPECTING THE NEW AGE

After all this research, we are left with a basic question; what does it mean to us? By every ancient reckoning, including our own Astrology, as well as the Mayans, the Hopi, the Aztecs, and many other indigenous peoples, we are indeed entering a New Age. This has become such a major belief in modern culture that it has created a new religion, New-Age Spirituality, which is a patchwork quilt of spiritual beliefs from many ancient traditions. The one thing they all have in common is the belief that we are on the verge of a new age of humanity, which will bring harmony, higher consciousness, and a spiritual awakening.

This new age is not necessarily safe either. As I mentioned at the outset of this book, there are a whole host of apocalyptic theories on how and when the earth may be destroyed. The myths and theories presented in this book are but a drop in the proverbial bucket. One of the popular ideas of the last decade is that we will likely be hit by an asteroid sooner or later. Such an event could be as cataclysmic for us as it was for the dinosaurs. And, as fate would have it, there are a number of near misses with asteroids predicted for the next few decades. If the scientists are off by as little as one degree here or there then we could well find that one of those asteroids is our doom. Then again, there is still thermonuclear war. Many Hopi shaman are convinced that this will be our downfall.

I have tried to use the guiding principle of science

throughout this book, even while studying some very outlandish theories and bizarre mythical accounts. But I cannot evaluate all the promises and changes of the New Age based upon science alone. Science gives us part of the story, but the rest is a matter of faith and intuition. I think if we are able to back away from our situation enough, we might just be able to see that the coming new age is actually much more simple and obvious than previously realized.

Never in the millions of years our species has lived and evolved on earth, as far as anyone knows, have our human ancestors flown through the air at the speed of sound itself, or communicated with people on the other side of the world, seeing them and hearing their voices in real time, through some tiny magic box, not to mention flying to the moon. To not appreciate that we are living a life far more amazing and miraculous than any ancient prophecy has foretold is the height of ignorance. We, here and now, are fulfilling ancient Vedic prophecies that talked of a golden age in the past and one that will be repeated in the distant future. The future is now. It's no coincidence that this time in history was predicted to be the start of a new age because we are turning that corner even as I write this.

Just look at the most miraculous of all the miracles of the modern age, the computer-internet revolution. We now have a global forum of consciousness and knowledge, not created by and for kings, but one built and populated by the people. Even more amazing is that anyone anywhere in the world can readily access the totality of recorded human knowledge in an instant, from a tiny device in their pocket.

We may use or abuse this power for the mundane, such as finding an all-night pizza delivery nearby, playing a videogame, downloading a YouTube video, finding an ex-girlfriend, or googling almost any question you may have from why is grass green, to how many liters are in a gallon? But we can also use this amazing miracle of technology to see any place on earth from space, in an instant. That's pretty cool. We can use it to find a cure for cancer, or make a breakthrough in understanding how to reverse global warming, or for trying to figure out when or if the world will end (my personal favorite).

For those who doubt that 2012 was the start of a new age, I wonder where they live, under a rock? Of course we are on the threshold of a new age. We just can't see it because we're too close to it, and because time moves so slowly to us. But, compare human culture during the last millennium up until the last 100 years, to life today. The contrast is astounding. We may not see it as a new age, but I guarantee you that archaeologists a thousand years in the future will! This is not unlike the Upper Paleolithic Revolution or the Neolithic Revolution.

We already have dubbed it the Industrial Revolution and the Information Age, but we are still looking at it too closely. We have to realize that one lifetime to us is everything, but to history it's a brief snippet of time. So it's hard for us to appreciate how dramatically or rapidly we have changed. We see the last 150 years as a long time. But in terms of archaeology, geology, and evolution, it's a mere blink of an eye. Thousands of years in the future, if humanity survives, they will be saying that sometime around 2000 AD (+/- 100 years) humanity suddenly transformed itself. And, while they may be able to trace the roots of this revolution back to discoveries and innovations of the last few hundred years, there will be no doubt that by the year 2012, human culture had fundamentally changed.

Polishing My Crystal Ball

If I have to put my prognosticator hat on and make predictions
about the future, I'd say that some of the features of the modern
internet culture, such the Arab spring, facebook, twitter, and their
use in politics, are just the very tip of a very large iceberg. I believe
that we truly do stand on the threshold of a new age. And, like
Oliver Reiser and many New Agers and even some modern- day
Maya Daykeepers, I see it as an age of global unity. Will it involve
higher consciousness? I don't see how it cannot. You cannot shine
such a bright light on everything in the world, every idea, every
scrap of evidence, every science, every voice, and allow everyone
in the world to see it, and yet somehow have the same or less
consciousness; that just doesn't make sense. This just may be the
start of what Einstein called Cosmic Humanism, by which he
meant a planetary consciousness.

Will it involve spiritual growth? Again I believe as Oliver
Reiser did that the evolution of humanity seems to be on a
trajectory of sorts, and the next step in the path seems to be one
of wisdom, and increased consciousness to the point of mystical
enlightenment. If we see the perfection of humanity as being
embodied in the life of Jesus, Buddha, or Gandhi, then yes, I'd
have to conclude that the future of humanity lies in completing our
spiritual development.

Of course I'd be quick to point out that statistically all of
us lie somewhere on a bell curve. And, at least half of all humanity
is statistically below average on any indicator you can think of. I
mean that, at any given point in evolution, there were always those
who were very smart and ahead of their time, and those who were

not so bright, whose genes were probably destined to become extinct. I've often speculated on the de-evolution of humanity due to my fear that the least wise and intelligent among us might end up having the most children. But we have to remember that at every point in evolution the bygones outnumbered the innovators at least a hundred to one.

How many people in the Renaissance studied the human anatomy of cadavers, or designed airplanes or submarines, as Leonardo da Vinci did? There was only one that we know of. How many people in the early 1900's were studying anatomy or designing airplanes, or submarines? Though they were still nowhere near a majority, there were quite a few more by then. How many people today study anatomy, or work in the aircraft manufacturing industry? The answer is millions! In just a few hundred years, the esoteric studies of da Vinci have become a downright common aspect of modern humanity.

Now, how many people today are capable of doing what it takes to become an astrophysicist, or a neuropsychologist? Not too many. So evolution moves like a large, slow herd, but it does progress. There is always a scout way up front, and a larger group of brave explorers behind them, and then there is always a huge bulk of humanity bringing up the rear.

The Myth of Progress

I know that the above scenario may rub some the wrong way. Many people in the last few decades, myself included, have commented on science's obsession with progress. They argue that our ancestors had a wisdom just as deep and profound as anyone today. Many might well argue that technological progress is not evidence of evolutionary progress, that this is a common myth. But I honestly think that this viewpoint has now become a bit irrelevant.

From everything I can see, the industrial and computer revolutions have already taken root and they have begun to change the world we live in. The next step in evolution must therefore be for our species to adapt to yet another environmental constraint, this time caused by technology. Exactly how this will happen is not yet clear. But, just as our Australopithecus ancestors would find it impossible to cope and survive in today's society, we can pretty much bet that the average person of today would find it equally hard to survive in a society of the distant future. This is evidence of a dynamic interface between mimetic and genetic evolution. As we mimetically change our world through technology, we will have to genetically evolve to adapt to our ever changing environment.

Given the above analysis, it is ironic to note that the next genetic mutation will likely come from the third world. This is because each time a baby is born in a technologically advanced culture, with some unknown neural structure in its brain, the abnormality is almost always removed in infancy. This is done before the doctors have any idea what this new structure might do, or what side-effects it may present for the child. I wish neurosurgeons would adopt a wait and see approach. But then again, if I put myself in the place of the parents, I would probably want some abnormal growth removed from my child's brain as well, just to be safe. And then there's the malpractice insurance companies, and I'm sure that they'd suggest erring on the side of caution as well. Nonetheless, it does make genetic mutations much less likely to take hold, if and when they do occur.

The Future of Humanity

You may well ask, what kind of adaptations will be required to live in a globally connected culture, made up of thousands of different languages, cultures, histories, traditions and beliefs? I think we will probably need to be more open-minded, more flexible, less rigid in our thinking and in our beliefs. We will likely become more intuitive, especially in human culture, just the way people today can spot an email scam within about 2 seconds. Reiser believed that this might include increased psychic abilities.

We will likely continue to be more of what we call humane, extending civil rights to all people, all over the planet, and even to animals, who many today believe are deserving of greater protection from exploitation and cruelty. We may even become a vegetarian species. And who knows? Maybe someday we will have plant rights as well, and begin harvesting fruit and vegetables in the most humane way possible. As we do this, we may well transform our consciousness, our perception, and increase our mental and emotional abilities as well. Though it may be wishful thinking, we would hope that crime, violence, and war would decline as well, as simple human decency increases.

As we change our social and built environment through technology, we continually must evolve to adapt to our environment. That's what evolution is. In this way the dynamic interaction of mimetic and genetic evolution has reached a tipping point. You might say that the culture and consciousness of humanity has caught fire! We are now altering our environment faster than evolution can help us to adapt. The other day I saw an 18 month old effectively using an iphone to play games, listen to music and even to call her grandmother. We may well see rapid evolutionary changes over the next millennium if not sooner. Again, I must stress that any of these changes will be focused on those few advanced scouts at first, while the rest of us bring up the rear. The brightest, most innovative, and most open-minded, flexible and intuitive thinkers will continue to push the envelope of innovation, and the rest of us will be forced to adapt, as we find the world around us changing.

By changing our environment through technology, we have sealed our fate. There is no doubt that, because of the technology that we already have, we are destined to create greater world unity, and greater tolerance and understanding of each other. This is just as Reiser predicted. The genes of those who cannot do so, either genetically or mimetically, will not survive the next few millennium.

The people that we see today as violent, belligerent or intolerant, such as terrorists, racists, and bigots will likely be seen as socially disabled thousands of years in the future. They simply will not be able to function in an open and global society. If they cannot be rehabilitated then they will likely be put in an assisted living facility and have their internet access limited. And so evolution will move on.

The same may well bode true for those with what psychologists call personality disorders. These are people who seem to have certain character defects, from birth. These can include traits such as a propensity for lying, cheating, stealing, paranoia, being antisocial, or a having a lack of emotion or empathy. These people may also someday be seen as severely disabled. They might be treated the same way we treat those with severe psychotic symptoms or dementia today. And again evolution will take its course.

The New Age is Now

On the whole I do believe in the most optimistic aspects of the ancient prophecies and New Age beliefs. As many of the Hopi elders have said, we are already there. We have already initiated the new age of humanity. 2012 was the conclusion to more than 100 years of dramatic changes. All we need now is the proverbial straw that broke the camel's back.

The littlest thing now could crystallize all the changes that we've already made, and force us to see the world anew. Perhaps it will be the discovery of extraterrestrial life, or the discovery of a jovian mass in the Oort cloud, or perhaps it will be a cataclysm or global disaster. No matter what happens, we will adapt. And that, by definition, is evolution.

So let's celebrate the New Age of humanity, and this New Age of Earth. Lets re-number our calendar, with next year being Year-One. Because it is here, now. We are the New Age! And it is up to us to determine the shape and quality of this brave new world.

Epilogue

PATHOLOGICAL SKEPTICISM: SEPARATING FACT FROM TRUTH

There have been a lot of myths, theories, and facts presented in this book. Some of the inferences that you can draw from this research are so outlandish as to be unthinkable to a trained scientist. And I'll be the first to admit that much of the evidence presented in this book is not yet properly vetted by the required rigorous research studies needed to confirm early findings. Skeptics have already objected to some of the findings presented here. And much of the theories that I present have not yet been accepted by mainstream science. So, I think it's worth examining the question of what constitutes good science.

Evidence is neither true nor false; it is the *conclusions* you draw from the evidence that are either true of false. Of course that depends on whom you talk to. One scientist's brilliant find is another scientist's mistake. Sometimes there is a consensus, but sometimes scientists are split. And sometimes the majority of scientists turn out to be wrong. Therein lies the difference between fact and truth. To understand this, let's go back and look at how we define errors in the scientific method.

First of all, there are two types of errors in science. There is a type one error, and a type two error. A type one error is the kind of error most people think of when they think of scientific errors. That is, a type one error is being too gullible, believing in crazy theories without sufficient proof. By this, you would think that the more skeptical you are, the less likely you are to be in error. But that is not always true.

A type two error can be just as bad as a type one error. A type two error is when you disbelieve something that is actually true. Remember Galileo and the Catholic Church. Well, his crazy theory turned out to be right, and they were wrong, but they refused to see that. They committed a huge type two error. Or how about that intelligence officer who warned his superiors that terrorists were planning on using airplanes as bombs to crash them into buildings? There was no great response to this threat. None of the security measures that were enacted after 9/11 were put into place at that time. Why? It was simply a type two error. And it was probably one of the worst in history.

Thalidomide

In real science, type two errors are every bit as dangerous as type one errors. In fact, in the field of foods and drugs, type two errors are much worse than type one errors. Think about it. Suppose they create a new drug to prevent morning sickness in pregnant women. If the scientists are wrong, and the drug doesn't prevent morning sickness, then the only harm done is to the company's profits. But, let's say that there is a theory that this drug might create birth defects. But there is no evidence to prove that it would create birth defects. So let's say that the hard-core skeptics won out, and they went ahead and released the drug, which had been proven to prevent morning sickness, but which had not proven to create birth defects. Well, this did happen. The drug was Thalidomide and guess what? This turned out to be a type two error. Many thousands of children were born horribly deformed because the drug company did not take seriously the unproven theoretical risk of birth defects. This has been called one of the biggest medical tragedies in modern times.

There are many other examples. There is the risk of nuclear reactors melting down, the risk of hexavalent chromium contaminating the ground water (the case that Erin Brockovich made famous), the risk of using mercury as a preservative in inoculations (still debated even though the drug companies have taken it off the market), the risk of carbon emissions creating global climate changes. All of these have been hotly debated, and some are still being debated. But every time a catastrophe happens, such as the nuclear accidents at Chernobyl, Three Mile Island, and Fukushima, we always find that hindsight is 20/20, and it was a type two error that was to blame. This goes for the Challenger disaster as well. Anytime someone is warned of a potential problem, and they fail to heed the warning because of a lack of proof, they are running the risk of a costly type two error.

So how does this happen? How can an otherwise rational, hard-nosed, tough-minded, skeptical scientist be so foolish that they do not to take into account a possible worst-case scenario, and plan for it accordingly? Good question. That is where the idea of *pathological skepticism* comes in. This is a concept that I came up with years ago to try to understand why some scientists are so skeptical that they frequently make type two errors, and often use wild leaps of illogic to support their erroneous conclusions. Since then, the term pathological skepticism has become something of a household word. Like so many psychological disorders, this disorder consists not of a conscious act but rather an unconscious thought process.

The Meditation Study

Skeptics often use straw-man arguments. What's funny about this is that they usually don't even know that they are doing it. Whenever I think of pathological skepticism, I always think of a psychological study done years ago at my Alma Mater, Long Beach State University. A researcher there, who shall remain nameless, wanted to study the effects of eastern meditation. So he came up with a pretty clever study. He came up with three options. A) One group practiced traditional meditation, as it is practiced in eastern religions. B) One group practiced a secularized Western form of meditation (essentially duplicating the religious meditation, without the religion). And, C) one group did nothing at all; this was the control group. He randomly assigned students to one of these three groups. He then monitored the subjects to look for positive outcomes, over time.

After the study was over, he looked at all the data. It clearly showed that both the eastern religion meditation group and the secular meditation group enjoyed positive benefits, and the control group did not. Both of the meditation groups were more relaxed, less stressed, and happier than the group that did not meditate. So, most people would say that he scientifically proved the effectiveness of meditation.

The funny thing was that he concluded that meditation did *not* work! You see he did not see the secular form of meditation as a form of meditation at all. He thought of it as a placebo. That is, like a sugar pill, an inert exercise. Even though he had duplicated many of the salient psychological features of eastern meditation, and perfectly distilled them in to an effective meditation practice, he saw it as nothing, just a placebo.

When he compared eastern meditation to his "placebo" meditation group and found no difference between the two groups, he concluded that meditation on the whole was apparently not effective. So this researcher not only proved the effectiveness of eastern meditation, he even came up with his own secularized version, easily practiced by millions of Americans, with all the same benefits of eastern meditation. And he didn't even realize what he had done! Ironically, years later Jon Kabat-Zinn would do essentially the same thing with his secular, mindfulness-based stress reduction (MBSR) practice (similar to the above "placebo") and he would become famous for it.

This researcher's mistake was completely unconscious and based on the tenacity of his beliefs. He believed that eastern meditation was fake, and that there was no positive results that came from practicing it. That is what he believed before he collected the evidence. Therefore, once he had the evidence, he interpreted it in the only way that made sense to him. In the end, he saw precisely what he wanted to see. His bias had eclipsed his insight, and he failed to see what he had actually discovered, which is that meditation does produce definite benefits and that it is a psychological process, independent of religion. You see when the facts conflicted with truth, he reinterpreted the facts to fit the truth. But, whose truth? Obviously, his truth is different from the truth of the Dali Lama or Jon Kabat-Zinn, or millions of people who meditate.

There is one final point to make here. When I first read this study I thought the researcher was purposely creating a straw-man argument. People do this in politics all the time. You mischaracterize your opponent's position, make it sound ridiculous, then you can easily argue against it. I thought this researcher was intentionally mischaracterizing eastern meditation as some kind of religious miracle that only benefits you if you are a devout Buddhist or something. In this way he could easily discredit eastern meditation by showing that it was no miracle of religion. Eventually I realized that this was indeed like a straw-man argument, but I don't think it was intentional.

Here the researcher really did believe that meditation was some kind of mystical religious experience that could only occur with divine intervention. In this way he was even more gullible than most people who actually practice meditation. He never even considered that it might simply be a scientific phenomenon. He thought he was striking a blow against superstition and ignorance, but he was only fighting shadows in his own mind. This is a classic case of what I call pathological skepticism.

Pathological Skepticism

I see this all the time with skeptics such as Michael Shermer, founding publisher of *Skeptic* magazine. Once, in a televised interview, I saw him comment on the Baghdad Battery, an ancient artifact found in an archaeological dig near Baghdad. The object was a terracotta jar, which had a copper cylinder inside of the jar, and an iron post that fit inside the copper cylinder without touching it. The metal was held in place in the jar with bitumen. It is believed that the jar had been filled with lemon juice or some other liquid, high in citric acid. Well, that sounds like a battery to anyone who knows about electricity. Shermer claimed that this artifact was not evidence that the ancients knew about electricity. He thinks that the device was merely used for electroplating, since we have found evidence of gold plated jewelry from that period. So, they did not know about electrical technology, they only knew about electroplating technology? I know what he means, but I don't think he realizes how obtuse that sounds. He just conceded that they did in fact know about the technology of electroplating, and actually knew how to build devices to implement it, including a battery cell sufficient to generate the necessary charge.

What he meant to say was that they did not have electrical wires suspended from electrical poles, with giant generators, producing massive energy to power light bulbs and run various appliances. But, he forgets that neither did Benjamin Franklin, Georg Ohm, or James Clerk Maxwell. Neither did Thomas Edison in the beginning. But surely he would not deny that these men knew of electricity. Shermer, in his defense, might say that even if they somehow knew about electroplating, then it still does not prove that they had any modern understanding of electricity. That's fair enough. I have no idea how they explained or thought about batteries or electricity. They may have attributed it to the gods, for all I know. But if you define a battery as any device that holds an electrical charge, and you define electricity as that energy, which is stored in a battery, then he is clearly wrong. They had obviously discovered the phenomenon that *we* call electricity. And if they did use it for electroplating, then that proves that they had discovered the relevant principles and materials involved in that process.

You see he's actually using a straw-man argument, and I don't think he's doing it intentionally. He thinks that when a crypto-archaeologist says that ancient civilizations "knew about electricity" or were "experimenting with it" that they're claiming that the ancients might have had electric lights and appliances, or that they lived just as we do today. But that's actually a straw-man argument. If he did this consciously, he would probably be doing it because he can't refute the actual facts so he misrepresents it as something crazy and then argues against that instead. The thing is, I don't think he even knows he's doing it. The crazy scenario he is arguing against is all in his head. Most people who are fascinated with the Baghdad battery don't really believe that the ancients who created it had electric toasters and florescent light bulbs.

This process of unconsciously creating straw-man arguments is really an example of a neurotic behavior, in that it is an unconscious process that skews a person's thinking and behavior often in illogical ways. So what would cause someone to do this? Unconsciously, this kind of person probably has a fear of being gullible himself, and doesn't want to be made a fool of by believing in a practical joke, or a con. So he dismisses, downplays, or distorts facts because they are in conflict with the *truth* as he knows it. I call this pathological skepticism, because it is basically a neurosis, a disorder that causes people to make errors in judgment, due to an unconscious fear of being humiliated or being overly gullible.

I've heard many such statements by various skeptics over the years. A good example of pathological skepticism is when someone says "there's no proof that was a UFO, it was probably just some flying object that just hasn't been identified yet." Or how about this one, "faith healing is nothing more than a placebo effect." And when you ask them what a placebo effect is, they respond, "a placebo is when the person is healed because they *believed* that they'd be healed." But isn't that another way of saying that they were healed because they had faith they'd be healed? These statements kind of remind me of the famous quote by Leo Durocher that, "anyone who goes to a psychiatrist ought to have his head examined." Just like the above skeptics, you know what he means but what he says is so obtuse that it's actually comical.

These kind of statements reveal the common fallacy of many skeptics' arguments. Let's take the statement on UFO's from above. "There's no proof that was a UFO, it was probably just some flying object that just hasn't been identified yet." This statement says it all. You know what the person meant to say, that there is no proof that it was an alien spacecraft, piloted by aliens. But, that's a straw-man argument. In most cases people don't say, "I saw an alien spacecraft." They usually just report seeing a UFO. If you're honest, then you'd have to admit that nobody really knows what they are; that's why they're called unidentified. So, in trying to refute the straw-man thesis (aliens), the skeptic accidentally acknowledges that there was in fact a valid report of an unidentified flying object (UFO). Likewise, when Shermer says that an ancient artifact is not evidence of electrical technology, rather it's only evidence of electroplating technology, he is acknowledging the very thing he is trying to refute. You see the pattern? It's like saying "that's not some woman, that's my wife."

The straw-man argument is actually an unconscious fear on the part of the skeptic. Often this is a fear that people may be saying something other than what they are actually saying. So, they hear "UFOs" but they think they heard "aliens." It may also be a neurotic need for closure. There are many scientific phenomena, such as unsolved UFO cases, for which there is no resolution yet.

This is usually just because the facts simply cannot be explained given the information that we currently have. Psychologically this can be very disturbing for some people. So they feel the need for closure. They want to believe that eventually it will be discovered to be a hoax. But until that happens they may experience a certain level of anxiety. So they will go ahead and assume it's a hoax, and even try to convince others that it's a hoax, thereby getting a feeling of closure. Then there is no pesky mystery hanging over their head, and they can sleep well at night.

The problem is that pathological skepticism can actually inhibit our ability to study and understand scientific phenomena. For instance, are scientists less likely to study paranormal phenomena, for fear of being ridiculed? Yes, of course! I have talked to many respected researchers and professors about their interest in paranormal phenomena at parties and behind closed doors. But because of the ridicule heaped on anyone who dares mention the word *paranormal*, scientists can't even openly discuss this topic. That is, unless they can prove that it is all a hoax, then the skeptical community breathes a collective sigh of relief, and their research is held up as a shining example of good scholarship. Sadly, when pathological skepticism inhibits scientific research, major mistakes can be made.

Piltdown Man vs. Australopithecus

Pathological skepticism is probably more common in the field of archaeology and anthropology than in any other field. Case in point, just look at Piltdown Man. There was a huge debate in anthropology at the beginning of the 20th century about whether our human ancestors first learned to walk then developed larger brains, or visa versa. Like many of his colleagues, one famous British archaeologist, Charles Dawson, a member of the Geological Society, and the Society of Antiquaries of London, believed that large brains came first, then the rest of the body evolved as it did. And in 1912 he found fossilized evidence to prove it. He presented a fossil showing the enlarged skull cavity of a modern human ancestor, but clearly it still had the jawbone of an ape, this fossil became known as Piltdown Man.

This new finding proved that the larger brain developed first. Finally, the matter was solved. Thanks to the discovery of Piltdown Man, there was no longer any mystery about which came first, big brains or walking upright. Despite some minor discrepancies in the fossil, this new view of human evolution was accepted by mainstream science.

Some years later, Raymond Dart, an unknown anatomist and researcher in South Africa, claimed that he found fossilized evidence of the opposite. It was essentially an ape that walked upright. He called his find *Australopithecus Africanus* or "Southern Ape Man from Africa."

So who was right and who was wrong? Well, it was clear from the start that the distinguished London archaeologist Dawson was right and this new fossil was irrelevant, and possibly a fake. After all, years earlier Dawson had already solved this riddle of evolution and proven for all time that our enlarged brains developed first. They all concluded that this unknown researcher, Dart, must be wrong. His find was dismissed as almost laughable. What he was claiming simply could not exist.

It took an additional 30 years before it was finally discovered that it was really Dawson's fossil that was a fake.

Piltdown man was nothing more than the cranium of a human with the jawbone of an Orangutan. It is believed that either he or one of his cohorts had created the fake fossil in order to prove his theory that an enlarged brain had preceded other advances in evolution. Then it was another 20 years before a more complete skeleton of an Australopithecus was discovered, proving that Raymond Dart was right all along. The new fossil was called *Australopithecus Afarensis,* and is better known as *Lucy.*

Some skeptics may say that they were too gullible and not skeptical enough in accepting the Piltdown Man and rejecting Australopithecus. But, alas, hindsight is 20/20. Think about it. Why did the skeptical scientific community believe a lie, and disbelieve bona fide evidence? It is because the lie was simply confirmation of an already accepted scientific theory. And the new evidence of Australopithecus contradicted what they thought they already knew to be true. The new evidence simply was not logical, given what they already *knew* about human evolution. Regarding Raymond Dart's find, they were not being too gullible; they were being too skeptical. When they rejected Australopithecus, they clearly had made a type II error.

Of course in hindsight we can see that they had actually decided which theory was correct first, then they looked for evidence that fit their beliefs. When contradictory evidence surfaced, they rejected and even ridiculed it. They did this not because of any serious study of the fossils, but rather because it did not fit with their beliefs.

Another telltale sign that this is pathological skepticism comes from the relative status of the two individuals involved. Skeptics are more likely to consider the source of the information than to actually look at the facts themselves. They eagerly believed the eminent scientist and were unduly suspicious of the unknown, less accomplished scientist, who they probably believed was just trying to attain fame and fortune, or possibly even trying to make fools of the scientific community.

That is not the way science is supposed to work. This sounds more like religion than science. Science should not be a matter of *belief.* It should be a search for facts, using the scientific method. Then you need to sift through possible explanations, and when you eliminate the impossible, whatever remains, no matter how improbable, is likely the most accurate conclusion.

In a line from the movie *The Last Crusade*, the intrepid archaeologist Indiana Jones says "Archaeology is the search for fact, not truth. If it's truth you're interested in, Professor Tyree's philosophy class is right down the hall." If only that were always the case, but as we've seen even archaeologists are not immune from confusing fact and truth.

Crypto-Archaeology

Currently, there is a bit of a revolution in the field of archaeology. There is a new breed of renegade archaeologists who are studying topics that are not usually accepted in the field of archaeology. While you may find papers published on these controversial topics, such as finding the historical basis for the myth of Atlantis, most main-stream archaeologists and skeptics scoff at such research. Why? This is because much of this new evidence goes against what mainstream archaeologists *believe* to be true.

Again, we see how scientists sometimes establish a belief system then look for evidence to support it. And any evidence that does not support it is discounted, downplayed, argued against, or claimed to be a fake, often using straw-man arguments. Sometimes, new theories such as Hapgood's earth crust displacement theory are just simply ignored. The result of this is that we are told to ignore any facts that do not fit with what mainstream science *believes* to be true. The irony is that this approach undermines the entire premise of science in the first place, which should favor facts over our preconceived notions of truth.

Speaking Fact to Power

A major problem that I see in recent years is that the entire topic of the Mayan prophecies of Doom have been pretty much banned in mainstream science. You can discuss it only if you are debunking it. If you are not debunking it, if you are really studying it seriously as we have in this book, then it is often dismissed as junk science. This is why it is important to separate truth from fact.

A good example of this is the information that I presented from Gavin Menzies' book *The Lost Empire of Atlantis*. One day, I needed to check the spelling of Menzies name, and I didn't have my books and papers handy, so I went online. I quickly found the name Gavin Menzies on a number of websites. But was that the same man? I opened the first website on the list; it was Wikipedia. I read the first paragraph description of him. I had to laugh when I read that regarding the Minoans "mainstream historians dismiss Menzies' theories and assertions as fictitious."

The entry said it so matter-of-factly that you'd swear that this book, which had just come out the previous year, had been thoroughly investigated and proven to be completely fraudulent. Even prior to reading his book, I had studied the Minoan culture, and had even gone to most of the same archaeological sites that he'd been to. So I already knew most of it was factual. And, I had also checked the sources for the new research that he presentd. At the very least, he has presented bona fide research from serious scholars, and woven these findings into a compelling theory. How can that be called fictitious? It's not even a new theory really. The similarities between the Minoans and the myth of Atlantis were recognized at least 40 years ago. Clearly, the so-called "mainstream historians" only call it fictitious because the *facts* that he presents do not fit the *truth* that they believe. So, it's just a matter of belief? That's not the basis of science; that's religion.

The bottom line is this. You cannot adequately make up your mind about something as momentous as a potential global cataclysm without taking into account these factors of pathological skepticism and our fallible human nature that causes us to see what we want to see and believe what we want to believe. The skeptics have always been there, criticizing anything that challenges our current beliefs. Only this time, it may be too important to let it slide. This time we may need to heed the warning of a potentially catastrophic theory, even if the definitive proof is still lacking. After all, if you find some puzzling facts that do not fit with what we know in mainstream science, then you don't stop researching it! In fact, that's a very good reason to research it further. It is usually in situations like this in which science either clears up confounding variables or makes breakthroughs.

We as a society just may need to wake up out of our complacent stupor and start holding the scientific community more accountable, and demand a reduction in type II errors, especially where our health and safety is concerned. This time we may not be able to afford to sit idly by and watch the proverbial Nero fiddling while Rome is burning. It's not just the threat of a global cataclysm, it is nuclear power plants that we are told are safe until disaster strikes, or global warming, which half of the population thinks is a hoax. Half of the population? How could that be? Clearly, if nothing else, this is an example of pathological skepticism run amok. Then there are the potential dangers of off-shore drilling, the storage of nuclear waste, reduced biodiversity in our food supply, genetic engineering, and the list goes on and on. These are all issues for which being too skeptical could kill us, or at least make life miserable for many people.

We claim to be a society that values science, but we actually value truth more than facts. But truth, alas, is whatever you believe it is. It is time that we as a society learned the difference between fact and truth. We as a people, and especially those of us in the scientific community, need to get back to the basics and start dealing with some inconvenient facts.

Conclusion

Finally, we conclude this book in a discussion of the excessive skepticism in the scientific community. We analyzed this phenomenon in relation to the scientific method, and mainstream science seems to come up lacking. In general, there appears to be a bias in the scientific community in which they are fanatically predisposed to guard against type I errors (being overly gullible), but consequently they make an excessive amount of type II errors (being overly skeptical). That's ok with inconsequential matters. But, as we have shown, if there is a potential threat to humanity, then a type II error can be disastrous. I found this is especially true with the study of a potential global cataclysm.

I call on all scientists to give up such shortsighted bias, and begin to study potentially important however unlikely theories with the same zeal that they apply towards making minor contributions to an existing body of knowledge, on topics that are almost always less consequential than a global cataclysm. In the words of F. N. Earll, regarding Hapgood's theory, "if it is an unworthy thing let it be properly destroyed; if not, let it receive the nourishment that it deserves."

And perhaps, just perhaps, we may want to at least start thinking about how to build an ark, …a really, really big ark.

REFERENCES

With some exceptions, all of the references refer to items listed in the bibliography section. Such exceptions include a full citation.

Chapter 1

1. Jenkins' *Maya Cosmogenesis 2012: The True Meaning of the Maya Calendar End-Date* .

2. Argüelés' *The Mayan Factor: Path Beyond Technology.*

3. Stuart's book *Order of Days.*

4. Aveni's book *The End of Time: The Maya Mystery of 2012*, or his article 'Apocalypse Soon?' In *Archaeology.*

5. See Haynes' *American Megafaunal Extinctions at the End of the Pleistocene.* Also see Haraldur Sigurdsson's paper in the journal *Eos* vol. 87, (34) August 22, 2006, in which he concludes that the destruction of Thera and Crete most likely lead to the myth of Atlantis. For more on the destruction of Crete and Thera, see Menzies' *The Lost Empire of Atlantis.*

6. Hapgood's *Path of the Pole, and* Cruttenden's book *The Lost Star of Myth and Time.*

7. Nibiru is an ancient Sumerian myth about a planet in our solar system that disappeared but is predicted to return one day. For a description of Nibiru, see WikiInfo on Nibiru (mythology).

8. For a good biography of Albert Einstein, see Walter Issacson's book *Einstein: His Life and Universe.*

9. The quotes presented here are not separately footnoted. For the most complete list of quotes by Einstein, see *The Quotable Einstein,* by Albert Einstein, compiled and edited by Alice Calaprice, and also see *Bite-Size Einstein: Quotations on Just About Everything from the Greatest Mind of the Twentieth Century,* by Albert Einstein, compiled and edited by Jerry Mayer, and John Holmes.

10. For more discussion on the topic of the skeptical mainstream scientific community, see the Epilogue to this book, Pathological Skepticism: Separating Fact From Truth

Chapter 2

1. Christenson's *Popol Vuh: The Sacred Book of the Maya*, and Goetz et al's *Popol Vuh: The Sacred Book of the Ancient Quiché Maya*.

2. Makemson's *The Book of the Jaguar Priest: A Translation of the Book of Chilam Balam of Tizimin with Commentary*, and Roys' *The Book of Chalam Balam of Chumayel*.

3. All of the quotes and references to the Popol Vuh in this chapter are taken from Christenson's *Popol Vuh: The Sacred Book of the Maya*, and Goetz et al's *Popol Vuh: The Sacred Book of the Ancient Quiché Maya*. For expediency, and due to the very large number of quotes used, the specific translator and page numbers are not listed for each quote. For more information on the Popol Vuh, please see these two excellent translations.

4. Stuart's *The Order of Days*.

5. Ibid.

6. Christenson's *Popol Vuh: The Sacred Book of the Maya*, and Goetz et al's *Popol Vuh: The Sacred Book of the Ancient Quiché Maya*.

7. *The Secret Tao: Uncovering the Hidden History and Meaning of Lao Tzu*, by D. W. Kreger.

8. Graham Hancock's *Fingerprints of the Gods: The Evidence of Earth's Lost Civilizations*.

9. Makemson's *The Book of the Jaguar Priest: A Translation of the Book of Chilam Balam of Tizimin with Commentary*, and Roys' *The Book of Chalam Balam of Chumayel*.

10. *Codex Perex and the Book of Chilam Balam of Mani* (1979), edited by E. Craine, and R. Reindorp, published by University of Oklahoma Press.

11. Roys' *The Book of Chalam Balam of Chumayel*.

12. Makemson's *The Book of the Jaguar Priest: A Translation of the Book of Chilam Balam of Tizimin with Commentary*. For expediency, and due to the very large number of quotes used, the specific page numbers are not listed for each quote. For more information see Makemson's book.

13. See Roys' *The Book of Chalam Balam of Chumayel*. For expediency, and due to the very large number of quotes used, the specific page numbers are not listed for each quote. For more information see Roy's book.

14. See Stuart's *The Order of Days.*

15. Ibid, and also see the works of Heley, Stray, and many others who have weighed in on the Tortuguero stele.

16. Van Stone's *2012: Science and Prophecy of the Ancient Maya.*

17. Stuart's *The Order of Days.*

18. Heley's *The Everything Guide to 2012.*

19. Shearer's *Beneath The Moon, and Under The Sun.*

20. Eberl & Prager (2005) p. 29-30, in Karl Herbert Mayer's *Maya Monuments: Sculptures of Unknown Provenance*, Supplement 4. Acoma Books.

21. Hunbatz Men's *Secrets of Mayan Science/Religion.*

22. Barrios' *The Book of Destiny: Unlocking the Secrets of the Ancient Mayans and the Prophecy of 2012.*

Chapter 3

1. Keber's *Codex Telleriano-Remensis: Ritual, Divination, and History in a Pictoral Aztec Manuscript.*

2. See Stuart's *The Order of Days.*

3. Ibid.

4. Hancock's *Fingerprints of the Gods: The Evidence of Earth's Lost Civilizations.*

5. Water's *The Book of the Hopi.*

6. 'Prophecies beyond 2012' regarding the Inca prophecies of Willaru Huayta, on the website www.alamongordo.com/prophecies_beyon_2012.html.

7. 'Maori Prophecy: Veil Dissolves in 2012' on the website: http://worldprophecies.net/Maori_Prophecy_Veil_Dissolves_in_2012.html.

8. Cruttenden's *The Lost Star of Myth and Time.*

9. Yukteswar's *The Holy Science.*

10. Ibid., p. 7.

11. Scranton's *The Science of the Dogon: Decoding the African Mystery Tradition.*

12. '#80: Zulu Prophecy: The Return of Mu-sho-sho-no-no,' on the website: diagnosis2012.co.uk/new.htm#zulu.

Chapter 4

1. *Plutarch Lives* (2010), published by The Folio Society, London.
2. Plato's *Timaeus and Critias*, p. 11.
3. Ibid., p. 10.
4. Ibid., p. 10.
5. Ibid., P. 10.
6. Plato's *The Statesman.*

Chapter 5

1. From the German translation of a cuneiform tablet describing Nibiru, by Wolfram von Soden in: *Zeitschrift für Assyriologie (ZA)*, no. 47, p. 17. See WikiInfo, Nibiru (mythology).
2. Sandars' *The Epic of Gilgamesh: An English Translation* (1960), by Penguin Classics.
3. See Genesis 6-10, in *The Holy Bible, King James Version*; and also see the *The Book* (1971) pp.6-9, by Tyndale House Publisher, Wheaton Illinois.
4. Hancock's *Fingerprints of the Gods: The Evidence of Earth's Lost Civilizations.*
5. Ibid.

Chapter 6

1. Cruttenden's *The Lost Star of Myth and Time.*
2. Aveni's book *The End of Time: The Maya Mystery of 2012*, and Stuart's *The Order of Days.*
3. Jenkins' *Maya Cosmogenesis: The True Meaning of the Maya Calendar End-Date.*
4. Cruttenden's *The Lost Star of Myth and Time.*
5. Ibid., p. 107.
6. Ibid., p. 108.

7. Ibid., pp. 108-109.

8. Ibid., pp. 121-122.

9. Ibid.

10. Yukteswar's *The Holy Science.*

11. See Max Born's *Einstein's Theory of Relativity*, p. 348, for a table of the observed and calculated values for the precession of Mercury.

Chapter 7

1. See Hancock's *Fingerprints of the Gods*, and the 2009 film '*2012*', directed by Roland Emmerich, which was based on Hancock's book.

2. Hapgood's *Path of the Pole.*

3. Ibid.

4. See Einstein's letter to Hapgood, in Hapgood's *Path of the Pole*, p. 327.

5. Einstein's Forward in *Hapgood's Path of the Pole*, p. xiv.

6. In Hapgood's *Path of the Pole*, p. 46.

7. Ibid.

8. Einstein's Forward in *Hapgood's Path of the Pole*, p. xiv.

9. Hapgood's *Path of the Pole*, p. 322.

10. Ibid.

11. Barbiero's article, 'On the Possibility of Instantaneous Shifts of the Poles' in *Lost Knowledge of the Ancients: A Graham Hancock Reader*, edited by Kreisburg, p. 202.

12. Ibid., p. 205.

13. Hapgood's *Path of the Pole.*

Chapter 8

1. Boyd's *Binary Star Hypothesis*, p.15

2. Ibid., p. 16.

3. See the Antoniadis et al paper: 'A Massive Pulsar in a Compact Relativistic Binary' in the April 26, 2013 issue of the journal *Science*.

4. Cruttenden's *The Lost Star of Myth and Time*, p. 129.

5. Ibid.

6. Ibid.

7. Ibid., p.130.

8. See 'Binary Neutron Star Collision' by David Bock, from the NCSA Visualization and Virtual Environments Group, on Haydenplanetarium.org/resources/ava/stars/S0606neutcoll.

9. Cruttenden's *The Lost Star of Myth and Time*, p. 174.

10. Schwaller de Lubicz's *Sacred Science: The King of Pharaonic Theocracy*.

11. Yukteswar's *The Holy Science*.

Chapter 9

1. See Reiser's books *Cosmic Humanism, The Integration of Knowledge,* and *A New Earth and a New Humanity*. Or for a very concise summary of Reiser's theories, see Heley's *Everything Guide to 2012*.

2. Ibid.

3. Ibid.

4. Reiser's *Integration of Knowledge*, p. 460.

5. Ibid., p. 454.

6. Ibid., p. 460.

7. Ibid., pp.458-460.

8. According to Heley's *Everything Guide to 2012*, p. 252.

9. Reiser's *Cosmic Humanism: A Theory of the Eight-Dimensional Cosmos Based On Integrative Principles From Science, Religion, and Art*, p. 531.

10. Ibid., p. 532.

11. Ibid., p. 533.

12. Heley's *Everything Guide to 2012*, pp. 253-254.

13. Ibid.

Chapter 10

1. See Andrew Gregory's Introduction in Plato's *Timaeus & Critias*, p. lv.

2. Plato's *Timaeus & Critias*, p. 113.

3. Angelos Galanopoulos was the first to claim that the myth of Atlantis was based upon the Minoan civilization and the apparent 'sinking' of their main city on the island of Thera (Santorini). For more information see Galanopoulos's *Atlantis: The Truth behind the Legend* (1969), by Bobbs-Merrill Co.

4. There are many articles and books on this topic. One of the more credible geological reports comes from Haraldur Sigurdsson's paper in the journal *Eos* vol. 87, (34) August 22, 2006, in which he concludes that the destruction of Thera and Crete most likely lead to the myth of Atlantis. For new archeological evidence of the Minoan's empire in the Atlantic, see the article on the excavations at Los Millares on the website, Ancient-wisdom.co.uk/spainlosmillares.htm. Also see W. Shepard Laird's article 'New Archaeological Evidence for Atlantis,' on http://newsgroups.derkeiler.com/Archive/Soc/soc.culture.europe/2007-06/msg00037.html

5. For more on Sir Arthur Evans, see Joseph Alexander MacGillivray's, *Minotaur: Sir Arthur Evans and the Archaeology of the Minoan Myth* (2000), New York: Hill and Wang

6. Christos G. Doumas' *Thera: Pompeii of the Ancient Agean: Excavations at Akrotiri 1967-1979*, (1983), Thames & Hudson.

7. Menzies' *The Lost Empire of Atlantis: History's Greatest Mystery Revealed.*

8. See the paper by University of Ulm pathologist, Svetlana Balabanova: 'First Identification of drugs in Egyptian Mummies' (1992), in *Naturwissenschaften*, vol 79, (8), pp. 358-358.

9. For more on this, see Brendan Burke's 'Materialization of Mycenaean Ideology and the Ayia Triada Sarcophagus' (2005), *American Journal of Archaeology*, vol. 109 (3), pp. 403-422.

10. For more, see Constantinos Triantafyllidis' The DNA of the Inhabitants of Greece, posted on *AntrhoScape*, http://s1.zetaboards.com/anthroscape/topic/4046419/1/. Also see his references including, Semino et al.'s 'Origin, diffusion and differentiation of Y-chromosome haplogroups E and J: inferences on the Neolithization of Europe and later migratory events in the Mediterranean area.' (2004), *American*

Journal of Human Genetics, vol. 74 p. 1023.

11. See Reidla et al's 'Origin and Diffusion of mtDNA Haplogroup X' (2003), *American Journal of Human Genetics* vol. 73 (5), p. 1023. Also see Virginia Morell's paper 'Genes May Link Ancient Eurasians, Native Americans.', *Science* vol. 280, p. 520.

12. For a concise description of the history and archaeology of the Old Copper culture of the Lake Superior Region, see 'The Old Copper Complex: North America's First Metal Miners & Metal Artisans,' online at copperculture.homestead.com/index.html

13. Drier, and Du Temple's *Prehistoric Copper Mining in the Lake Superior Region* (1961), Calumet, MI: R.W.Drier.

14. See Menzies' *The Lost Empire of Atlantis: History's Greatest Mystery Revealed,* pp. 296-298, and see color plates pp.230-231.

15. Ibid, p. 290. For more on this topic see BYU professor emeritus of anthropology, John L. Sorenson's 'Ancient Voyages Acrss the Ocean to America: No Longer Impossible,' posted online at www.bmaf.or/node/200, or anyone of his books such as *An Ancient American Setting for the Book of Mormon* (1985), *Transocianic Culture Contacts between the Old and New Worlds in Pre-Columbian Times: A Comprehensive Annotated Bibliography* (1988), or *World Trade and Biological Exchanges before 1492* (2004).

16. See *BBC News Online,* Tom Housden's article 'Lost City Could Rewrite History,' January 19, 2002, posted at http://news.bbc.co.uk/2/hi/south_asia/1768109.stm

17. Badrinaryan's article 'The Gulf of Khambhat: Does the Cradle of Ancient Civilization Lie off the Coast of India?' in *Lost Knowledge of the Ancients: A Graham Hancock Reader,* edited by Kreisburg.

18. Wenke's *Patterns in Prehistory: Mankind's First Three Million Years.*

19. Ibid.

20. Badrinaryan's article 'The Gulf of Khambhat: Does the Cradle of Ancient Civilization Lie off the Coast of India?' in *Lost Knowledge of the Ancients: A Graham Hancock Reader,* edited by Kreisburg, p. 166.

21. A detailed criticism of the finds from the bay of Khambhat (Cambay) can by found in Paul V. Heinrich's article 'Artifacts or Geofact? Alternative Interpretations of Items from the Gulf of Cambay,' posted on www.intersurf.com/˜chalcedony/geofact.html. Also, a brief synopsis of the evidence and arguments for and against this potential archaeological find are found in the article 'Marine Archaeology

in the Gulf of Cambay' posted on www.reference.com/browse/Cambay%2C+Gulf+of

22. Wenke's *Patterns in Prehistory: Mankind's First Three Million Years*.

23. Woof's *A Short History of the World: The Story of Mankind from Prehistory to the Modern Day*.

24. Hancock's *Fingerprints of the Gods: The Evidence of Earth's Lost Civilizations*.

25. For more information see Daniel D. Lukenbill's *Ancient Records of Assyria and Babylonia* (1989), by Histories & Mysteries of Man.

26. *The Secret Tao: Uncovering the Hidden History and Meaning of Lao Tzu*, by D. W. Kreger.

27. Hapgood's *Maps of Ancient Sea Kings*.

28. See Riane Eisler's *The Chalice and the Blade: Our History, Our Future* (1988), New York: Harper Row.

29. Rose's article 'New Light on Human Prehistory in the Arabo-Persian Gulf Oasis'.

30. Ibid.

31. Lawler's article 'The World in Between: 5,000 Years Ago, A Long-Buried Society in the Iranian Desert Helped Shape the First Urban Age'.

32. For more information see Haywood's *Historical Atlas of the Ancient World: 4,000,000 – 500 BC*, and King's *Atlas of Human Migration*.

33. See Romey's Diving the Maya Underworld, *Archaeology*, (2004) vol. 57 (3)

34. See D. Rebikoff's Underwater archeology: Photogrammetry of artifacts near Bimini.' (1979), *Explorers Journal*. vol. 57, (3), pp. 122-125. Also see J. A. Gifford's The Bimini 'cyclopean complex (1973), *International Journal of Nautical Archaeology and Underwater Exploration*. vol. 2, (1), p. 189. And, for a more speculative reading see Dennis Brook's *Atlantis was America: Tampa was the Royal City*, (2008), BooksSurge Publishing.

35. See *Daily Mail* article 'The 10,000 Year Old Boy's Bones Found in an Underwater Mexican Cave That Could Rewrite the History of the Americas' on www.dailymail.co.uk/sciencetech/article-1305929/Ancient-skeleton-prehistoric-child-removed-Mexican-underwater-cave.html

36. 'The Fuente Magna of Pokotia Bolivia,' posted on www.factulty.ucr.edu/˜legneref/archeol/fuentema.htm

37. See Arthur Faram's 'The Yonaguni Pyramid: A Geoglyphic Study of the Yonaguni Monolith, Japan' posted on www.yonaguni.ws. Also see Graham Hancock's website: www.gramhancock.com/gallery/underwater/Yonaguni.htm

38. Haynes' *American Megafaunal Extinctions at the End of the Pleistocene.*

39. King's 'Glaciology & Geodynamics.'

40. Hapgood's *Path of the Pole.*

41. Ibid., p. 276.

42. Ibid.

43. Swaminathan's 'Neolithic Community Centers at Wadi Faynan, Jordan.'

44. Barkai & Liran's 'Midsummer Sunset at Neolithic Jericho.'

45. Scham's 'The World's First Temple: Turkey's 12,000 Year-Old Stone Circles Were the Spiritual Center of a Nomadic People'.

46. Marija Gimbutas' *Goddesses and Gods of Old Europe: Myths and Cult Images* (1974), by Thames and Hudson, and also *The Civilization of the Goddess: The World of Old Europe* (1992), Harper Collins.

47. Scham's 'The World's First Temple: Turkey's 12,000 Year-Old Stone Circles Were the Spiritual Center of a Nomadic People'.

48. Constantinos Triantafyllidis' The DNA of the Inhabitants of Greece, posted on *AntrhoScape,* http://s1.zetaboards.com/anthroscape/topic/4046419/1/. And also see his references including, Semino et al.'s 'Origin, diffusion and differentiation of Y-chromosome haplogroups E and J: inferences on the Neolithization of Europe and Later Migratory Events in the Mediterranean Area.' (2004), *American Journal of Human Genetics* vol. 74 p. 1023

49. Hapgood's *Path of the Pole.*

50. Ibid.

51. Hancock's *Fingerprints of the Gods: The Evidence of Earth's Lost Civilizations.*

52. Hapgood's *Path of the Pole*, p. 286.

53. Hancock's *Fingerprints of the Gods: The Evidence of Earth's Lost Civilizations*, p. 89.

54. Ibid.

55. Doore's *Markawasi: Peru's Inexplicable Stone Forest.*

56. Ibid.

57. Haynes' *American Megafaunal Extinctions at the End of the Pleistocene*, and also see extensive research from Hapgood's *Path of the Pole*.

58. Ibid.

Chapter 11

1. Hapgood's *Maps of Ancient Sea Kings*.

2. Hapgood's *Path of the Pole*.

3. Ibid., p. 4.

4. Ibid., p. 5.

5. Ibid., p. 7.

6. Ibid., p. 8.

7. Ibid., pp. 6-7.

8. Ibid., pp. 8-9.

9. Ibid., pp. 9-10.

10. Ibid., p. 14.

11. See Haywood's book *Historical Atlas of the Ancient World, 400,000 - 500 BC*. Though focused on the history of human habitation, he presents maps that show where the ice sheets were during the last ice age, p. 1.02., as well as how water levels rose, from the end of the last ice age to the present.

12. Ibid., pp. 94-95.

13. Ibid., p. 108.

14. Ibid., p. 121.

15. Martínez-Frias et al's 'A Review of the Contributions of Albert Einstein to Earth Science.'

16. Woelfli & Baltensperger's 'A Link Between an Ice Age Era and a Rapid Polar Shift.'

17. Ibid.

18. Warlow's 'Geomagnetic Reversals?'

19. Slabinski's 'A Dynamic Objection to the Inversion of the Earth on its Spin Axis.'

20. Ibid.

21. Steinberger & Torsvik's 'Absolute Plate Motions and True Polar Wander in the Absence of Hotspot Tracks.'

22. King's 'Glaciology & Geodynamics.'

23. Ibid., p. 433.

24. Ibid.

25. See Krause's criticism of Hapgood's Theory on his website: www.skrause.org/writing/papers/Hapgood_and_ecd.shtml.

26. Eden's 'The Polar Shift: A New Look at Earth's Changing Past.'

27. Barbiero's 'On the Possibility of Instantaneous Shifts of the Poles.'

28. Earl's Forward to Hapgood's *Path of the Pole*, pp. vii-ix.

29. Ibid.

30. Ibid.

31. Ibid.

32. Ibid., p. ix.

33. Ibid.

34. Mather's Forward to Hapgood's *Path of the Pole*, p. xii.

35. Ibid., p. xiii.

Chapter 12

1. Cruttenden's *The Lost Star of Myth and Time*

2. Ibid.

3. See Matese & Whitmire's 'Persistent Evidence of a Jovian Mass Solar Companion in the Oort Cloud.'

4. Ibid., p. 931

5. Ibid., p. 931.

6. Ibid.

7. Lissauer et al's 'A Jovian Mass Object in the Oort Cloud?'

8. the Cosmos TV Online article 'Has a New Plantet (Planet X or Tyche) Been Discovered in Our Solar System?' *Cosmostv.org*, July 5, 2011.

9. Ibid.

10. Matese & Whitmire's 'Persistent Evidence of a Jovian Mass Solar Companion in the Oort Cloud.,' p.926.

11. Matson's 'What's Flinging Comets Out of the Oort Cloud?'

12. Chapter 5 of Crutenden's *The Lost Star of Myth and Time*, pp. 143-176.

13. Schwaller de Lubicz's *Sacred Science: The King of Pharaonic Theocracy.'*

14. See May 20, 2008 article by Jeff Hecht, in NewScientist: Space, "Gravity Probe B Scores 'F' in NASA Review."

15. Ibid.

Chapter 13

1. Oliver Reiser wrote a number of books. Three of them seem especially relevant and those are *A New Earth and A New Humanity* (1942), *The Integration of Human Knowledge* (1958), and the culmination of his life's work in *Cosmic Humanism: A Theory of the Eight Dimensional Cosmos Based on Integrative Principles from Science, Religion, and Art.* (1966)

2. Carl Sagan's *The Cosmic Connection: An Extraterrestrial Perspective.*

3. Collin's *The Cygnus Mystery.*

4. See Phil Plait's article 'No A Pole Shift Won't Cause Global Superstorms,' on the *Discover Magazine* website in the Bad Astronomy section: February 9th 2011. http://blogs.discovermagazine.com/badastronomy/2011/02/09

5. Collin's *The Cygnus Mystery.*

6. Bar-Yosef's 'The Upper Paleolithic Revolution.'

7. Simmons' *The Neolithic Revolution in the Near East: Transforming the Human Landscape.*

8. Sham's 'The World's First Temple.'

9. Barkai & Liran's 'Midsunner Sunset at Neolithic Jericho,' and Swaminathan's 'Neolithic Community Centers at Wadi Faynan, Jordan.'

10. Sham's 'The World's First Temple.'

11. Shoch's book *Voices of the Rocks*, as well as his follow-up article 'New Studies Confirm Very Old Sphinx.

12. Robert Bauval's book *The Egypt Code*, as well as his article 'The Egypt Code: Is the Key to Egypt's Past Reflected in the Stars Above Giza?' in *Lost Knowledge of the Ancients: A Graham Hancock Reader*, edited by Kreisburg, pp. 30-37.

13. Hancock's *Fingerprints of the Gods: The Evidence of Earth's Lost Civilizations*, p. 89.

14. Hapgood's *Path of the Pole*.

15. There are numerous references mentioned previously. See Bar-Yosef's 'The Upper Paleolithic Revolution,' Simmons' *The Neolithic Revolution in the Near East*, Sham's 'The World's First Temple,' Barkai & Liran's 'Midsunner Sunset at Neolithic Jericho,' and Swaminathan's 'Neolithic Community Centers at Wadi Faynan, Jordan,' just to name a few.

16. Haynes' *American Megafaunal Extinctions at the End of the Pleistocene*.

17. Hapgood's *Path of the Pole*.

18. Simmons' *The Neolithic Revolution in the Near East: Transforming the Human Landscape*.

19. Refer back to Plait's 'No A Pole Shift Won't Global Superstorms,' cited earlier in this chapter.

SELECTED BIBLIOGRAPHY

Aveni, Anthony (2009). *The End of Time: The Maya Mystery of 2012*. Boulder, CO: University of Colorado Press.

Aveni, Anthony (2009). 'Apocalypse Soon?: What the Maya Calendar Really Tells Us About 2012 and the End of Time.' *Archaeology*, November-December, p. 31-35.

Argüelés, Jose (1987). *The Mayan Factor: Path Beyond Technology*. Rochester, VT:

Bear & Co.

Argüelés, Jose (1988). *Earth Ascending: An Illustrated Treatise on the Law Governing Whole Systems. Rochester*, VT: Bear & Co.

Boyd, George S. (1937). *A New Theory of the Solar System: A Binary Star Hypothesis*. Kingsport, TN.

Barbiero, Flavio (2010). 'On the Possibility of Instantaneous Shifts of the Poles: Has It Occurred During Human Existence?' In Kreisberg, Glenn (Ed.), *Lost Knowledge of the Ancients: A Graham Hancock Reader*. Rochester VT: Bear & Co.

Barkai, R. & Liran R. (2008). 'Midsummer Sunset at Neolithic Jericho.' *Time & Mind: The Journal of Archaeology, Consciousness, and Culture*, 1/3, p.273-284.

Barrios, Carlos (2009). *The Book of Destiny: Unlocking the Secrets of the Ancient Mayans and the Prophecy of 2012*. New York: HarperOne.

Bar-Yosef, Ofer (2002). 'The Upper Paleolithic Revolution.' *Annual Review of Anthropology*, 31: 363-93

Bauval, Robert (2010). The Egypt Code. New York: The Disinformation Comapany.

Benedict, Gerald (2010). *The Maya 2012: The End of the World or the Dawn of Enlightenment?* London: Watkins Publishing.

Benedict, Gerald (2008). *The Mayan Prophecies for 2012*. London: Watkins Publishing.

Boyd, George S. (1937). *A New Theory of the Solar System: A Binary Star Hypothesis*. Kingsport, TN.

Christenson, Allen J. (2007). *Popol Vuh: The Sacred Book of the Maya. The Great Classic of Central American Spirituality, Translated from the Original Maya Text*. Norman, OK: University of Oklahoma Press.

Collins, Andrew (2010). *The Cygnus Mystery*. London: Watkins Publishing.

Cosmos TV Online (2011). 'Has a New Planet (Planet X or Tyche) Been Discovered in Our Solar System?' *Cosmostv.org*, July 5, 2011.

Cruttenden, Walter (2006). *Lost Star of Time and Myth*. Pittsburgh, PA: St. Lynn's Press.

Curry, Andrew (2012). 'Ancient Germany's Metal Traders: A Post-Cold War Construction Boom is Exposing Evience of a Powerful Bronze Age Culture.' *Archaeology*, 65 (3) May-June, pp.30-33.

Doore, Kathy (2006). *Markawasi: Peru's Inexplicable Stone Forest*. Surprise, AZ: Kathleen Doore.

Earll, F. N. (1970). 'Forward to the Second Edition.' In Hapgood, Charles H. *Path of the Pole*. Kempton, IL: Adventures Unlimited Press.

Eden, Dan (2004) 'The Polar Shift: A New Look at Earth's Changing Past,' posted on *Viewzone.com*.

Einstein, Albert (1958). 'Forward to the First Edition.' In Hapgood, Charles H. *Path of the Pole*. Kempton, IL: Adventures Unlimited Press.

Einstein, Albert, and Calaprice, Alice (Ed.) (1996). *The Quotable Einstein*. Princeton, NJ: Princeton Univeristy Press.

Goetz, D., Morley, S., & Recinos, A., (1950). *Popol Vuh: The Sacred Book of the Ancient Quiché Maya*. Norman OK: University of Oklahoma Press.

Hancock, Graham (1995). *Fingerprints of the Gods: The Evidence of Earth's Lost Civilization*. New York: MJF Books.

Hapgood, Charles H. (1999). *Path of the Pole*. Kempton, IL: Adventures Unlimited Press.

Hapgood, Charles H. (1966). *Maps of Ancient Sea Kings: Evidence of Advanced Civilization in the Ice Age*. New York: Chilton Books.

Harada, Takehisa, & Shimura, Michiyoshi (1979). 'Horizontal Deformation of the Crust in Western Japan Revealed from First-Order Triangulation Carried Out Three Times.' *Tectonophysics*, 52, (1-4), p. 469-478.

Haynes, Gary (2009). *American Megafaunal Extinctions at the End of the Pleistocene (Vertebrate Paleobiology and Paleoanthropology Series)*. New York: Springer Science + Business Media LLC.

Haywood, John (2010). *Historical Atlas of the Ancient World: 4,000,000 – 500 BC*. New York: Metro Books.

Heley, Mark (2009). *The Everything Guide to 2012: All You Need to Know About The Theories, Beliefs, and History Surrounding the Ancient Mayan Prophecies*. Boston, MA: Adams Media.

Huggett, R. J. (1988). 'Terrestrial Catastrophism.' *Progress in Physical Geography*, 12, (4), p. 509-532.

Hutton, William, & Eagle, Jonathan (2004). *Earth's Catastrophic Past and Future: A Scientific Analysis of Information Channeled by Edgar Cayce*. Boca-Raton FL: Universal Pub.

Issacson, Walter (2007). *Einstein: His Life and Universe.* New York: Simon & Schuster.

Jenkins, John Major (1998). *Maya Cosmogenesis 2012: The True Meaning of the Maya Calendar End-Date.* Rochester, VT: Bear & Co.

Keber, Eloise Quiñones (1995) *Codex Telleriano-Remensis: Ritual, Divination, and History in a Pictoral Aztec Manuscript.* Austin, TX: University of Texas Press.

Kenyon, J. Douglas (2005). *Forbidden History: Prehistoric Technologies, Extraterrestrial Intervention, and the Suppressed Origin of Civilization.* Rochester VT: Bear & Co.

King, R. F. (1959). 'Glaciology and Geodynamics.' *Journal of Glaciology,* 4, (36), p. 432-434

King, Russel (Ed.) (2007). *Atlas of Human Migration.* Buffalo, NY: Firefly Books.

Kolata, Alan L. (1993). *The Tiwanaku: Portrait of an Andean Civilization.* Cambridge, MA: Blackwell Publishers.

Kreisberg, Glenn (2010). *Lost Knowledge of the Ancients: A Graham Hancock Reader.* Rochester VT: Bear & Co.

Kreger, D. W. (2014). *Documenting Evidence of a Native American Astronomical Marker.* Poster session presented at the Annual Meeting of the Archaeological Institute of America, Chicago, IL. Abstract published in the AIA 115th Annual Meeting Abstracts vol. 37, p. 81.

Kreger, D. W. (2012). *2012 & The Mayan Prophecy of Doom.* Palmdale, CA: Windham Everitt Publishing.

Kreger, D. W. (2011). *The Secret Tao: Uncovering the Hidden History and Meaning of Lao Tzu.* Palmdale, CA: Windham Everitt Publishing.

Kreger, D. W. (2010). *A Survey of Rock Art Used in Initiation Ceremonies: Further Explorations of the Neuropsychology of Neolithic Shamanic Imagery.* Poster session presented at the Annual Meeting of the Archaeological Institute of America, Anaheim, CA. Abstract published in the AIA 111th Annual Meeting Abstracts, vol. 33, p. 44.

Kreger, D. W. (2000). The Lost Temple of the Goddess. *Caribbean World Magazine*, issue 33, p. 66.

Lawler, Andrew (2012). 'Rethinking The Thundering Hordes: How Herding Nomads Created a Network That Carried Civilization Across Central Asia More than 4,000 Years Ago.' *Archaeology*, 65 (3) May-June, pp.42-47.

Lawler, Andrew (2012). 'The Pearl Trade: Archaeologists Excavating on the Shores of the Persian Gulf Search for What May Prove to be the Source of the World's Longest Lived Economy.' *Archaeology*, 65 (2) May-June, pp.46-51.

Lawler, Andrew (2011). 'The World in Between: 5,000 Years Ago, A Long-Buried Society in the Iranian Desert Helped Shape the First Urban Age.' *Archaeology*, 64 (6), Nov-Dec. pp. 24-31.

Lemesurier, Peter (2010). *Nostradamus, Bibliomancer: The Man, the Myth, the Truth.* Pomotn Plains, NJ: New Page Books.

Lissauer, Jack J., Matese, J. J., & Whitmire, D. P. (2011). *A Jovian Mass Object in the Oort Cloud?* Paper presented at the meeting of the American Astronomical Society, Boston, MA.

Makemson, Maud W. (1951). *The Book of the Jaguar Priest: A Translation of the Book of Chilam Balam of Tizimin, with Commentary.* New York: Henry Shuman.

Martínez-Frías, Jesús, Hochberg, David, & Rull, Fernando (2006). 'A Review of the Contributions of Albert Einstein to Earth Sciences-in commemoration of the World Year of Physics.' *Naturwissenschaften*, 93: 66-71.

Martin, Hugo (2011). 'Mexico Hopes Visitors Put It On The Calendar.' *Los Angeles Times*, Saturday, September 17th, Business Section.

Matese, John, & Whitmire, Daniel P. (2011). 'Persistent Evidence of a Jovian Mass Solar Companion in the Oort Cloud.' *Icarus*, Vol. 211, (2), pp. 926-938.

Mather, Kirtley F. (1959). 'Forward to the First Edition.' In Hapgood, Charles H. *Path of the Pole*. Kempton, IL: Adventures Unlimited Press.

Matson, John (2011). 'What's Flinging Comets Out of the Oort Cloud?' Transcript of a pod-cast by *Scientific American*, May 31, 2011.

McKenna, Terrance and Dennis (1994). *The Invisible Landscape: Mind, Hallucinogens, and the I Ching*. New York: HarperOne.

Men, Hunbatz (1990). *Secrets of Maya Science/Religion*. Rochester VT: Bear & Co.

Menzies, Gavin (2011). *The Lost Empire of Atlantis: History's Greatest Mystery Revealed*. New York: William Morrow / HarperCollins Pulishers.

Plato (2010). *Statesman* (Jowett, B., Trans.). Charleston SC: Nabu Press.

Plato (2008). *Timaeus and Critias*. (Waterfield, R., Trans.). New York: Oxford University Press.

Reiser, Oliver L. (1966). *Cosmic Humanism: A Theory of the Eight-Dimensional Cosmos Based on Integrative Principles from Science, Religion, and Art.* Cambridge, MA: Schenkman Publishing Co.

Reiser, Oliver L. (1958). *The Integration of Human Knowledge: A Study of the Formal Foundations and the Social Implications of Unified Science.* Boston, MA: Extended Horizons Books.

Reiser, Oliver L. (1942). *A New Earth and a New Humanity.* New York: Creative Age Press.

Rose, Jeffrey I. (2010) 'New Light on Human Prehistory in the Arabo-Persian Gulf Oasis', *Current Anthropology*, 51:849–883.

Roys, Ralph L. (1967). *The Book of Chilam Balam of Chumayel.* Norman, OK: University of Oklahoma Press.

Sagan, Carl (1975). *The Cosmic Connection: An Extraterrestrial Perspective.* New York: Dell Publishing Co.

Saturno, W., Stuart, D., Aveni, A., & Rossi, F. (2012). Ancient Maya Astronomical Tables from Xultun, Guatemala. *Science*, 336 (6082) pp. 714-717.

Scham, Sandra (2008). 'The World's First Temple: Turkey's 12,000 Year-Old Stone Circles Were the Spiritual Center of a Nomadic People.' *Archaeology*, 61, (6) p. 22-27.

Schoch, Robert M. (2005). 'New Studies Confirm Very Old Sphinx: Orthodox Protests Notwithstanding, Evidence for the Schoch/West Thesis is Growing.' In **Kenyon, J. Douglas** (2005). *Forbidden History: Prehistoric Technologies, Extraterrestrial Intervention, and the Suppressed Origin of Civilization.* Rochester VT: Bear & Co., pp. 95-100.

Schoch, Robert M. (1999). *Voices of the Rocks*. New York: Harmony Books.

Schilken, Regis (2009). 'Book Review: Path of the Pole by Charles Hapgood.' *Blogcritics.org*, March 7th.

Schwaller de Lubicz, R. A. (1982). *Sacred Science: The King of Pharaonic Theocracy*. Rochester, VT: Inner Traditions.

Scranton, Laird (2006). *The Science of the Dogon: Decoding the African Mystery Tradition*. Rochester VT: Inner Traditions.

Shearer, Tony (1975). *Beneath The Moon and Under The Sun*. Santa Fe, NM: Sun Publishing Co.

Shryock, Andrew, & Smail, Daniel L. (2011). *Deep History: The Architecture of Past and Present*. Berkeley, CA: University of California Press.

Simmons, Alan H. (2011). *The Neolithic Revolution in the Near East: Transforming the Human Landscape*. Tucson, AZ: University of Arizona Press.

Slabinski, V. J. (1981). 'A Dynamic Objection to the Inversion of the Earth on its Spin Axis.' *Journal of Physics A: Math, General*, 14, 2503-7.

Steinberger, Bernhard & Torsvik, Trond (2008). 'Absolute Plate Motions and True Polar Wander in the Absence of Hotspot Tracks.' *Nature*, 452, pp. 620-623.

Stray, Geoff (2005). *Beyond 2012: Catastrophe or Awakening?* Rochester VT: Bear & Co.

Stuart, David (2011). *The Order of Days: The Maya World and the Truth About 2012*. New York: Harmony Books.

Swaminathan, Nikhil (2012). Neolithic Community Centers at Wadi Faynan, Jordan, in 'Top Ten Excavations of 2011.' *Archaeology*, 65 (1) p.26.

Symmes, Patrick (2010). 'History in the Remaking: A Temple Complex in Turkey that Predates Even the Pyramids is Rewriting the Story of Human Evolution. *Newsweek*, March 1st.

Timms, Moira (2005). 'Destination Galactic Center: John Major Jenkins Thinks Today's World Has Much to Learn from the Ancient Maya.' In Kenyon, J. Douglas (2005). *Forbidden History: Prehistoric Technologies, Extraterrestrial Intervention, and the Suppressed Origin of Civilization*. Rochester VT: Bear & Co., pp. 124-130.

Van Stone, Mark (2010). 2012: *Science and Prophecy of the Ancient Maya*. Imperial Beach, CA: Tlacaelel Press.

Voorhies, Barbara (2012). 'Games Ancient People Played: An Intriguing Discovery in a Mexican Swamp Provides Evidence of Earliest Form of Amusement in the Americas.' *Archaeology*, 65 (3) May-June, pp.48-51.

Warlow, P. (1978). 'Geomagnetic Reversals?' *Journal of Physics A: Math, General*, 11, 2107-2130.

Waters, F. (1963). *The Book of the Hopi*. New York: Balentine.

Wenke, Robert J. (1980). *Patterns in Prehistory: Mankind's First Three Million Years*. New York: Oxford University Press.

Woelfli, W, & Baltensperger, W (2004). 'A Link Between an Ice Age Era and a Rapid Polar Shift.' Unpublished paper in *Cornell University Library Database*, arXiv: physics/0407082.

Woolf, Alex (2008). *A Short History of the World: The Story of Mankind from Prehistory to the Modern Day*. New York: Metro Books.

Yukteswar, Swami Sri (1990). *The Holy Science*. Los Angeles: Self Realization Fellowship. (Originally published in 1894).

INDEX

ABOUT THE AUTHOR

D. W. KREGER

Dr. Kreger is a psychologist, an expert on the occult, and a researcher in the fields of psychology, archaeology, and ancient mysticism. He holds a Ph.D. in clinical psychology, completed his post-doctoral training in neuropsychology, and is a Diplomate of the International Academy of Behavioral Medicine, Counseling, and Psychotherapy. In addition to his psychological research, he has investigated archaeological sites in 17 countries around the world. His work has been presented at major academic conferences, and appeared in both research and popular media. He is the author of *The Secret Tao: Uncovering the Hidden History and Meaning of Lao Tzu*. Currently, he teaches psychology at the MacLauren Institute, and is a consulting clinical psychologist in private practice. He lives with his family on a small vineyard, north of Los Angeles, CA.

Made in United States
North Haven, CT
04 April 2023

34991530R00163